T0234314

A Ludic Journey into Geometric Topology

Ton Marar

A Ludic Journey into Geometric Topology

 Springer

Ton Marar ⓘD
ICMC
University of São Paulo at São Carlos
São Carlos, São Paulo, Brazil

Translation from the Portuguese language edition: Topologia Geométrica para Inquietos by
Ton Marar, © 2019 by Ton Marar. Published by Edusp – Editora da Universidade de São
Paulo. All Rights Reserved.

ISBN 978-3-031-07444-8   ISBN 978-3-031-07442-4   (eBook)
https://doi.org/10.1007/978-3-031-07442-4

Mathematics Subject Classification: 54-01, 51-01, 00A05

This Springer imprint is published by the registered company Springer Nature Switzerland AG
The registered company address is: Gewerbestrasse 11, 6330 Cham, Switzerland

To Agnaldo Aricê Caldas Farias, teacher

Foreword

In the history of Western thought, geometry goes hand in hand with philosophy. From a handful of techniques invented, according to Herodotus, around 1300 BC by the land surveyors of the Nile valley, over centuries it became the bridge between the world of ideas and the world of things. As a whole, this discipline is multifaceted, uncovering a spectrum of theories, all of which allude to the need to represent and study physical space in its most variegated aspects.

Ton Marar, the author of this precious text, leads us by the hand through the intricate paths of this ancient science, which is nowadays indispensable for understanding our universe. In these modern mirabilia, the inquisitive but curious reader will find many gems, such as the classification of Platonic solids and that of surfaces, the concept of orientability, and many other ideas and suggestions for future journeys into fascinating but certainly impervious territory.

São Carlos, Brazil   Igor Mencattini
February 2022

Preface

In the famous video game of the 1980s, the character Pacman moves on a rectangular screen in two perpendicular directions, down or up and left or right. Pacman's universe is two-dimensional, and he does not know what it is like to go forward or backward from the screen. Furthermore, when Pacman crosses the left edge of the screen, he appears at the same height on the right edge. Similarly, it happens when he crosses the horizontal edges.

Pacman's two-dimensional world is immersed in a surface without a boundary, an endless surface shaped like a donut. This surface is called a torus.

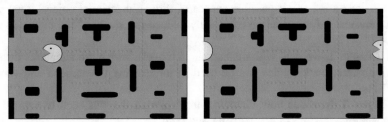

How would it be possible for Pacman to understand the toric shape of his world? For him, the third dimension is an esoteric space. He cannot get out of his two-dimensional world and enjoy it in three-dimensional space, like we do. Pacman is confined to his two-dimensional world and the only chance for him to understand the shape of his world is by deduction, an intellectual activity beyond physical sensation.

Our situation is not very different. We know little about the shape of the universe in which we live. Is our universe infinite? If it is finite, it must have no boundary; otherwise, what would be beyond the boundary?

There is no chance of leaving this universe to perceive its shape, as we did with the world of Pacman. We will have to deduce it.

To solve this great mystery, mathematics, or more specifically geometry, is fundamental.

It was easy to determine the shape of Pacman's universe. We just needed to gather some physical information from that world, which implied identifying the vertical edges and the horizontal edges of the rectangle where Pacman lives his life. Then, through an extra dimension, we made the identifications and finally saw the shape of Pacman's two-dimensional world without a boundary.

By analogy, from the three-dimensionality of our universe, it would take at least one extra dimension to be able to *see* its shape: a fourth dimension.

In this book, we are going to show the reader how to develop some sensitivity to see certain three-dimensional objects without boundary, which we call hypersurfaces.

Is our universe a hypersurface? Astrophysicists have the task of describing the cosmos geometrically, and some of them believe that the universe is modeled by a hypersurface, as shown in the article by JP Luminet et al. in Nature 425 (2003), *Dodecahedral space topology as an explanation for weak wide-angle temperature correlations in the cosmic microwave background.*

If one day the hypothesis is confirmed that our universe is in fact a hypersurface and we have a list of all possible shapes of three-dimensional objects, then with some physical information we can finally deduce the shape of the universe, who knows?

In Chapter 1, we deal with mathematical models, which are allegories through which abstract mathematics can be used in interpreting phenomena and solving problems.

Chapter 2 describes how Platonic and Keplerian theories seek to explain the cosmos with a mixture of mathematics and faith.

A brief account of geometries from Felix Klein's point of view is found in Chapter 3.

The next chapter contains an introduction to topology as a kind of geometry, and we present the classification of finite-sized, borderless two-dimensional objects called closed surfaces. These surfaces are divided into two classes, namely orientable and non-orientable. The first ones have two sides, they define an interior and an exterior, while the others, the non-orientable ones, all contain a Möbius strip, which is a truly one-sided surface.

There are several areas of knowledge that make use of the topological classification of closed surfaces. On October 3, 2016, the Nobel Prize in Physics was awarded to a trio of British materials experts. In their work entitled *Topological phase transitions and topological phases of matter,* certain exotic states of matter (beyond solid, liquid and gas) that occur at extreme temperatures are described. In this research, the authors used the topological classification of closed surfaces, and the practical result is a series of new superconducting materials.

From the classification of closed surfaces, we can see the difficulty of obtaining a topological classification of hypersurfaces. These geometric objects live in high-dimensional spaces, at least four, and this is why the fourth dimension is introduced in Chapter 5.

Three-dimensional models of non-orientable closed surfaces are covered in Chapter 6, and the final chapter contains some hypersurface models.

I would like to thank my various students from the mathematics for architecture discipline in the architecture and urbanism course at IAU-USP, São Carlos, Brazil, who for decades motivated me to write this text. Thanks are also due to many of my colleagues, among them Carlos Martins, David Sperling, and Marcelo Suzuki from IAU-USP; Tiago Pereira, Igor Mencattini, Farid Tari, and Ali Tahzibi from ICMC-USP; Flávio Coelho and Paolo Piccione from IME-USP; Stefano Luzzatto from ICTP, Trieste; and Robinson dos Santos and Martin Peters from Springer. The presentation of several chapters improved a great deal after Alessandra Pavesi carefully read the book, to whom I am eternally grateful.

São Carlos, Brazil   Ton Marar
February 2022

Contents

Chapter 1
Mathematical Models

Contrary to the belief that mathematics is a collection of problem-solving techniques, historian and mathematician Morris Kline (1908–1992) provides the following description: *Mathematics is more than a method, an art, and a language. It is a body of knowledge with content that serves the physical and social scientist, the philosopher, the logician, and the artist; content that influences the doctrines of statesmen and theologians; content that satisfies the curiosity of the man who surveys the heavens and the man who muses on the sweetness of musical sounds; and content that has undeniably, if sometimes imperceptibly, shaped the course of modern history* [2, p. 9].

This body of knowledge has been under construction for millennia. Mathematicians participate in this process mainly driven by curiosity, like a climber who climbs a mountain because it is there, although practical questions have also motivated the theoretical development of mathematics from the beginning.

The activity of researchers in mathematics is purely intellectual and, in principle, it is not related to our physical world, as everything happens in the perfect world of ideas. Their conclusions, in the form of theorems, are defended with logical arguments to convince non-believers. The set of arguments is called the proof of the theorem. Often a proof does not stop the curiosity of finding a new one, for the same result, more objective, more beautiful. Beauty that can be quantified by the minimality of arguments.

How can it be that mathematics, being after all a product of thought, independent of experience, is so admirably adapted to the objects of reality? [1] asked the perplexed Albert Einstein (1879–1955).

Using mathematics in everyday situations takes place mainly through the so-called mathematical models, that is, allegories that adapt the real problem to the world of ideas, creating the possibility of dealing with problems scientifically.

Models have accompanied us since childhood and some even entertain us. A broomstick once played the role of a horse. In this case, we have a physical model, which is not very close to a real horse.

© The Author(s), under exclusive license to Springer Nature Switzerland AG 2022

T. Marar, *A Ludic Journey into Geometric Topology*,

https://doi.org/10.1007/978-3-031-07442-4_1

In addition to physical models, there are conceptual models. For example, a solid sphere is the locus of points in three-dimensional space whose distances to a given point (the center) are less than or equal to a given measure (the radius). Thus, the solid sphere is conceptually conceived. The adjective solid is used to distinguish this three-dimensional object from the two-dimensional sphere, which is just the surface, the shell, of the solid sphere (surface, therefore without thickness, therefore two-dimensional).

If, on the one hand, a marble is a physical model of a solid sphere, on the other hand, a solid sphere is a conceptual model of the marble. When the principles that define a conceptual model are mathematically based, then the model is called a mathematical model.

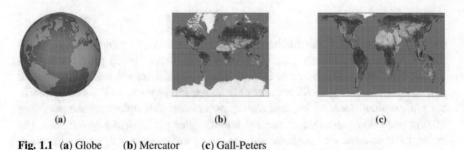

(a)   (b)   (c)

Fig. 1.1 (a) Globe   (b) Mercator   (c) Gall-Peters

A globe (Fig. 1.1a) is a physical model of our planet, which also has a solid sphere as a mathematical model. The sphere conceived by mathematicians is a perfectly homogeneous geometric object, which does not exist in nature and has characteristics quite different from those of our planet. Seen from afar, the planet Earth may even look like a sphere, but we, who live on its surface, with rivers and mountains, know that this planet is far from being a sphere. However, for certain purposes, the sphere is a suitable model for the Earth, even for aesthetic reasons. The perfection of the geometric object sphere pleases us and, perhaps for this reason, it is considered a good model for our planet since antiquity.

A world map, which is obtained from a projection of the globe onto a sheet of paper, is a representation of the planet best suited to be included between the pages of a book. Different projections generate different designs; the most popular being the Mercator projection (Fig. 1.1b). This projection distorts countries affecting territorial areas; Alaska seems to have the same area as Brazil, when in fact Brazil is five times larger. Another projection, called the Gall-Peters projection (Fig. 1.1c), takes into account the territorial area of the countries and has a strange aspect, as we are used to the Mercator projection.

In early 2017, after the election of the US president, a series of untruths was reported, which received the kind name of fake news. Contrary to untruths, the Boston City Department of Education decided to indicate the Gall-Peters projection for its public schools to develop a study of geography closer to reality, at least as far as the territorial area is concerned. Perhaps this marks the beginning of the end of the nearly 500-year reign of the Mercator projection, which deludes us about the relative areas of countries, just as we were misled one day that a broomstick was a horse.

Modeling, also called representation, creates an association between an abstract space and a physical space. This passage from the abstract to the concrete, and vice versa requires certain agreements, which are the subject of profound logical-philosophical discussions.

Consider the case of representing a plane of the Euclidean geometry. A sheet of paper serves as a model of a portion of that two-dimensional space, for its thickness is negligible compared to its width and length. We must understand that, as in abstract two-dimensional space, the sheet is made up of infinite points. When we mark two points on this sheet with a pen and draw the straight line segment between the two points, we are just distinguishing the points in the abstract space that make up the line segment. A circle in the plane is the selection of points equidistant from a fixed point. The inside of the circle is a set of points that we call a disk. On the sheet, we draw a circle and fill in the interior to designate the disk. Abstract two-dimensional space, represented by sheets of paper, is the place where all lines and figures reside.

In addition to a good model of a Euclidean plane, on the sheet of paper, we can also represent objects that live in the third dimension. A cube, for example, can be represented on the sheet, but we have to agree with certain representational devices. In this case, we use perspective and dotted lines to denote the lines of the object hidden from the observer. But it was not always so; perspective only began to be used with geometric rigor around the fifteenth century.

Creating good representations; that is, those that have the main properties of what you want to represent and that are reasonably simple, is an art. This is well developed in mathematics, particularly geometry. Also in algebra, numerical representation has gone through several periods, until reaching the current symbology. Imagine if we still used Roman numerals instead of Arabic numerals; the number 888 would be expressed in Roman notation by twelve letters $DCCCLXXXVIII$.

Cartography stands out in terms of being careful with representation, as can be found in an exaggerated short story written by Jorge Luis Borges (1899–1986) entitled *Del rigor en la ciencia: In that empire, the art of cartography achieved such perfection that the map of a single province occupied the entire city, and the map of the empire, the entire province. Over time, these excessive maps did not satisfy and the College of Cartographers raised a map of the empire, which had the size of the empire and coincided punctually with it.* In 1982, Umberto Eco (1932–2016) describing details of Borges' short history, wrote the article *On the impossibility of drawing a map of the Empire on a scale of 1 to 1.* But, in 1893, much before Borges and Eco were born, Charles Dodgson (1832–1898), better known as Lewis Carroll, had published *Sylvie and Bruno Concluded*, where a map on the scale of a mile to the mile was also considered. It was never spread out as the farmers objected that the use of the map would cover the whole country and shut out the sunlight!

When making a mathematical model, care must be taken not to exaggerate the details. The more sophisticated; that is, the closer the created model is to the reality in question, the more complex the mathematics is needed to deal with it.

A notable example of mathematical modeling took place in England in 1667 when a ripe apple fell off the tree and hit a Cambridge University academic on the head. It is narrated that right after the accident, the subject hit by the ripe apple started to describe the movement of falling bodies. His theory, viewed with disbelief by the scientific community at the time, revolutionized our understanding of certain facts that occur near the surface of our planet, as well as some aspects of the cosmos. Certainly, many ripe apples must have hit the heads of inhabitants of that island, but it was on Isaac Newton's head (1643–1727) (Fig. 1.2a) that the phenomenon managed to produce so much knowledge.

(a)   **(b)**

Fig. 1.2 (**a**) Newton (**b**) Galileo

The first mathematical model offered by Newton to describe a falling body was quite simple: the entire mass of the body would be concentrated in a point (the Euclidean point of the pure mathematician) and would be subjected to a single force, which he called the force of gravity, which pulls the object towards the center of the Earth (spherical). It is known, however, that this model had already been formulated by Galileo Galilei (1564–1642) (Fig. 1.2b) decades before Newton. In fact, from Galileo's considerations we have what are now known as Newton's first and second laws of motion, respectively the law of inertia and $F = m\,a$. In his work on the motion of bodies, Newton acknowledged, albeit timidly, in Galileo the original author of these two laws [5, p. 21]. The law of inertia states that an object remains at rest or in uniform, unaccelerated motion unless an external force acts on the object. This law gives rise to the concept of mass. The second law states that force is equal to mass times acceleration.

With this simple model, the movement of a falling body at each instant of the fall can be described. In the case of the apple that, as it detaches itself from the tree, has an initial speed equal to zero, the speed v after t seconds would be, according to Galileo, $v = g\,t$, where g is a constant, called by Newton the acceleration of gravity. Therefore, the speed would increase over time proportionally to the constant g.

One of the problems with this simple model is that a heavy object, such as an anvil, and an apple would fall in the same way, in which the speed increases in proportion to the time of the fall. Intuitively this does not seem right: bodies of such different masses showing the same behavior when falling. The reason for this incongruity lies in the initial assertions in making the mathematical model, for example, the assertion that only a single constant force acts on the falling object. Although ill-set, the problem thus modeled provides a very simple solution. That is, even if the model represents the problem relatively poorly, the answer is quite straightforward and simple.

Newton created another, slightly more sophisticated model to describe falling bodies. He assumed that in addition to the force of gravity, which pulls the object down toward the center of the sphere that shapes the planet, another force, called air resistance, acted on the object in the opposite direction to the fall. When we put our hand out the window of a moving car, or even a wagon from Galileo's time, we experience air resistance. This is a great contribution from Newton's work on the motion of bodies, through which he introduces us to his third law, called the law of action and reaction.

Newton assumed, based on experience and intuition, that for bodies falling from a great height, the air resistance was proportional to the square of the velocity. Thus, his model would take into account the object's mass, in addition to being possible to consider characteristics of the falling body through the constant which multiply the velocity squared in the equation of the phenomenon. In this case we are faced with a model that appropriately discriminates between falling anvils and apples. It is a model closer to the reality of the phenomenon. However, the speed of the object, which in the previous model was simply g times t, is now expressed by the function $\tanh(x) = 1 - 2/(e^{2x} + 1)$ called the hyperbolic tangent:

$$v = \sqrt{\frac{mg}{k}} \tanh\left(\sqrt{\frac{kg}{m}}t\right)$$

In case the reader does not know the hyperbolic tangent function, to understand it, a little differential and integral calculus is needed, a discipline whose creation is attributed to Newton himself and also to the German Gottfried Leibniz (1646–1716), who both claimed paternity.

The model of the falling body under the action of gravity, in one direction, and the force of air resistance, in the opposite direction, is more faithful to the phenomenon. Every body that falls from a great height, like a parachutist, initially experiences an accelerated movement when the air resistance is still small, as the speed is low. However, as speed increases, so does the resistance of the air, until this force against the fall equals the force of gravity. From the exact moment when the forces of gravity and air resistance are equal, the accelerated falling body starts a fall in uniform rectilinear movement, therefore without acceleration, but with a large initial velocity, called *terminal velocity,* which remains until the parachute opens. In the case of the anvil and the apple, the terminal velocities are different, when dropped simultaneously from the same height, the apple reaches its terminal velocity much earlier than the anvil, so it takes longer to reach the ground.

Until Galileo, the physics introduced by Aristotle (384 BC–322 BC) reigned, which lacked rigor, but few would dare to question it. Some argue that Aristotle claimed that bodies fall with a speed proportional to their weight. Others, in Aristotle's defense, say he was referring to terminal velocity, which is actually affected by the weight of the body. *That Aristotle ever supposed for an instant that a 2-lb. weight fell, in the ordinary sense of words, twice as fast as a 1-lb. weight is an absurdity,* wrote Major John H. Hardcastle (1870–1937) [4].

The example of the falling body illustrates how it is possible to sophisticate a mathematical model that describes a given physical phenomenon, making it more and more faithful to the phenomenon, but with the consequent increase in the complexity of the solution. Ultimately, if we consider that the most diverse forces act on a falling body, the solution of an equation that faithfully corresponds to this reality would be so complex that it would be convenient to leave the task of describing the phenomenon to poets, instead of scientists. Thus, even if the result was not so useful from a scientific point of view, it would at least please the senses.

In the 1970s, mathematical models occupied space in important daily newspapers in various parts of the world, an unusual fact for mathematics, whose evolution takes place without revolutions and, therefore, without many chances of becoming news in daily newspaper. The headlines of that time announced a new theory, named *Catastrophe theory* by its creator, the Frenchman René Thom (1923–2002).

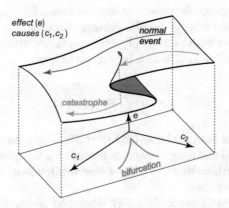

Fig. 1.3 Catastrophe

Until then, physical phenomena whose models were described by independent variables, let us call them *causes,* had a continuity relationship with the dependent variables, let us call them *effects*; that is, it was common to think that if the cause was continually varied, the effect would also vary continuously.

Catastrophe theory classifies several models in which continuous changes in cause can bring about abrupt changes in effect (Fig. 1.3), hence the name catastrophe. Such abrupt changes occur when the causes of the phenomenon reach values that belong to a particular set, called *the bifurcation set*, which does not exist for normal events.

Under the command of British mathematician Sir Christopher Zeeman (1925–2016), the mathematical models of Catastrophe theory were used in specialties as diverse as sociology, psychology, biology, linguistics, engineering and economics. They served to create a unifying and simplified mathematical discourse.

However, the relevance of this new language far from fulfilled the initial revolutionary promise, published in daily newspapers such as L'Express in France and The New York Times in the United States. These, in fact, publicized the wonders of the new theory by naming René Thom as the French Isaac Newton of the twentieth century and Catastrophe theory as the most important development in mathematics since the invention of differential and integral calculus 300 years earlier.

Furthermore, at the end of the twentieth century, a new theory, called Chaos theory, attracted a great deal of attention, even from laypeople, due to the beautiful figures produced by computers. This new theory would present mathematical models of evolutionary processes that occur in nature. With it also came the famous fractals (Fig. 1.4) that excite the imagination of many. However, just as Catastrophe theory, fractals receive severe criticism, particularly from philosophers of science who do not accept the attempt to relate them to natural phenomena.

Fig. 1.4 Fractal

Controversies aside, mathematical models are fundamental to the use of mathematics in interpreting natural problems, even if only approximately. Particularly in the case of problems involving dynamic processes, differential and integral calculus is an essential tool for solving mathematical models.

In the seventeenth century, the fatherhood of differential and integral calculus led Newton and Leibniz, the former English and the latter German, into a bitter intellectual dispute. What is certain is that both contributed enormously to our, however humble, understanding of the nature of the universe.

Averse to nationalist disputes, Einstein opened a lecture in 1929 at the University of Sorbonne in Paris with the following anecdote: *If my theory of relativity is proven correct, Germany will claim me as a German and France will declare that I am a citizen of the world. Should my theory prove untrue, France will say that I am a German and Germany will declare that I am a Jew* [3, p. 98].

In addition to modeling things, mathematics also serves to model situations, doubts and even delude us about predicting the future. St. Augustine (354–430) already warned us, according to Morris Kline: *The danger already exists that the mathematicians have made a covenant with the devil to darken the spirit and to confine man in the bonds of Hell* ([2], p. 3).

From the point of view of many students, traumatized by the set of techniques they are forced to memorize in mathematics classes around the world, Augustine was right. For this reason, beautiful mathematical models that have helped us to solve problems, even if approximate, can become important allies in combating this plague that keeps students away from mathematics.

References

1. Einstein, Albert ; *Geometrie und erfahrung*, Springer 1921. English translation https://mathshistory.st-andrews.ac.uk/Extras/Einstein_geometry/
2. Kline, Morris; *Mathematics in Western Culture*, Oxford University Press 1953.
3. Knowles, Elisabeth ed.; Oxford dictionary of modern quotations, Oxford University Press 2007.
4. Hardcastle, J.H.; *Professor Turner and Aristotle*, Nature 92, p. 584 (1914).
5. Newton, Isaac; *Mathematical Principles*, trad. Andrew Motte 1729, University of California Press 1974.

Chapter 2
Ancient Greek Big Bang Theory

2.1 Platonic Solids

In the beginning, there was only water and the heavens. Everything was empty until Tupanã came down in the midst of a great wind... According to an Amazonian legend, the deity embodied a vortex in order to create the universe.

Since then we have been in constant rotation at an astronomical speed. While the Earth takes 24 h to complete a rotation around its own axis, and one year to revolve around the Sun, the solar system as a whole takes over 200 million years to revolve around the axis of our galaxy, the Milky Way. The last time the Sun occupied its present place, dinosaurs reigned over our planet.

The speeds of these rotational movements are immense. On Earth, the rotation speed varies with latitude: near the equator, it is much greater than near the poles, as the distance to be covered to complete a rotation is greater. Thus, the Earth rotates around its axis with an equatorial speed of 1700 km per hour and, around the Sun, at a much faster speed, of more than 100,000 km per hour. The Solar System, in turn, revolves around the Milky Way, and this one around... Well, if we add only the known figures that make the cosmos a gigantic carousel, we reach the extraordinary speed of 700,000 km/h.

The more our knowledge of the universe broadens, the more this speed increases, and, paradoxically, our insignificance is confirmed. If one day we deluded ourselves that we occupied the center of the universe, today we would confront our condition of cosmic dust. But always in a rotational movement.

In the absence of a definitive theory about the creation of the universe, we are left with myths, legends and philosophical reflections. One of the earliest descriptions can be found in the dialogue *Timaeus* [8] by Plato (427 BC–347 BC) (Fig. 2.1a), the most restless of all Greek philosophers. Plato presents his version of the creation of the universe using the perfect and eternal world of forms as a model. Timaeus explains to Socrates (470 BC–399 BC) the origin of our physical world, which in the early days only existed in a state of disorder, using certain geometric allegories that were supposedly used by the demiurge—a geometer creator!

The idea of an architect of the universe appears with some frequency in the literature.

(a)   (b)   (c)

Fig. 2.1 (a) Plato   (b) Bible Moralisée   (c) Urizen

An old image is in the 1220 edition of the Bible (*Bible moralisée*). Another, published 570 years later, belongs to the mythological work *Europe: A Prophecy* by the visionary artist William Blake (1757–1827). In the image of the Bible, the creator is ready to set the world in rotation. The design is made using a compass. Within a circle are a spherical moon and sun, and in the center a shapeless matter. This matter will become Earth when God applies the same geometric principle to it (Fig. 2.1b).

Blake, on the other hand, shows Urizen with his compass open in the dark to impose a rational order on chaos (Fig. 2.1c). In both images, the Platonic inspiration is evident.

Among the geometric allegories, Timaeus highlights a set of five polyhedra, known today as Platonic solids. Four of these solids represent the fundamental elements—fire, earth, air and water, from which, according to Timaeus, everything derives. The fifth element represents the universe, the quintessence.

If today, on the one hand, we know that this is not so, on the other hand, we do not know whether particles such as leptons and quarks are, in fact, the fundamental elements. Plato paved the way for this quest after carrying out his seminal work. The Timaeus dialogue is the big bang theory of ancient Greece.

Platonic solids are three-dimensional convex objects whose boundaries, that is their surface, define regular polyhedra; in other words, they are composites of vertices, edges and faces, along with the three-dimensional region they enclose. The vertices must be points on a sphere that will circumscribe the solid. The edges are straight line segments of equal length, and the faces, bounded by edges, are regular polygons of the same type (equilateral triangles, squares, regular pentagons).

Furthermore, each vertex of the polyhedron has the same number of faces; that is, they are indistinguishable.

Theorem *There are only five types of platonic solids.*

It is difficult to know when this result was established. Renowned British authors, M. Atyah and P. Sutcliffe, in the article *Polyhedra in Physics, Chemistry and Geometry,* [1] express their belief that stones from the Neolithic period found by archaeologists in Scotland (Fig. 2.2) constitute evidence that the so-called Platonic solids were already known at least a thousand years before Plato.

However, these stones from the Neolithic period are just works of art that have little or nothing to do with the Platonic solids.

Certainly, a proof that only five types of polyhedra meet the conditions described appeared in ancient Greece, as the deductive method, necessary for the proof, was only established at that time!

Euclid (325 BC–265 BC) attributes the construction of three of the five solids to Pythagoras (570 BC–490 BC) and the other two to Theaetetus of Athens (415 BC–369 BC).

Fig. 2.2 Carved stone

Two of the five solids appear in ancient texts, long before Plato: they are the cube and the pyramid with a triangular base. The cube has six faces, which is why it is also called the hexahedron. On each of its eight vertices, there are exactly three of the six square faces. The pyramid with a triangular base has four faces and is therefore called the tetrahedron. On each of its four vertices there are exactly three of the four faces, which are equilateral triangles.

A proof that there are only five Platonic solids follows the argument recorded in the final propositions of the last of the 13 books of Euclid [2], a work that intended to record all the mathematical knowledge of the time. A relationship between edge of the boundary of the solid and the diameter of the surrounding sphere is also expressed there.

For example, (Fig. 2.3) in the case of the cube with side s, if the diameter of its surrounding sphere (i.e., the length of the main diagonal of the cube) measures d, then the edge of the inscribed tetrahedron (i.e., the diagonal of the face of the cube) measures $\sqrt{2}s$. These three segments are sides of a right triangle. Then, by the Pythagoras theorem, $d^2 = s^2 + (\sqrt{2}s)^2 = (3/2)(\sqrt{2}s)^2$.

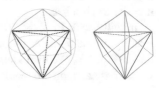

Fig. 2.3 Cube and tetrahedron

To show that, in addition to these two examples (cube and tetrahedron), there are only three other Platonic solids, we start by observing that the faces of these polyhedra are just equilateral triangles, squares or regular pentagons. In fact, from the convexity of the polyhedron, it follows that at each vertex, the sum of the angles of the incident polygonal faces must be less than 360°. Since at least three faces must fall on each vertex, regular hexagons or polygons with more than six edges would add up to 360° or more, as the interior angles of these polygons measure 120° or more.

We will describe all the possible Platonic solids in the following three steps.

Step 1:   Construction of possible platonic solids with triangular faces.

There are three types, namely, the tetrahedron, the octahedron (eight faces) and the icosahedron (twenty faces) (Fig. 2.4).

Fig. 2.4 Tetrahedron, octahedron and icosahedron

To see this, we start by assuming that at each vertex three triangular faces are incident and, therefore, the angles of the faces incident to each vertex add up to $3 \times 60 = 180°$. From these hypotheses, we obtain the tetrahedron.

Now, assuming that four triangular faces meet at each vertex, the angles of the faces incident at each vertex add up to $4 \times 60 = 240°$. From these hypotheses, we obtain the octahedron, whose model is obtained from two pyramids with a square base, identified by the bases.

Assuming that five triangular faces are incident at each vertex, the angles of the incident triangular faces at each vertex add up to $5 \times 60 = 300°$. From these hypotheses, we obtain the icosahedron.

This exhausts possible platonic solids with triangular faces. In fact, with six equilateral triangles falling on one vertex, the sum of the angles would reach 360°, and thus the construction would not be convex.

Among the three regular polyhedra with triangular faces, the icosahedron is the most elaborate. A model of the icosahedron can be built on paper and this will make it easier to visualize.

Initially, with 10 of the 20 triangular faces of the icosahedron, we created a band identifying the triangles by one of their edges, alternating the top vertices, up and down (Fig. 2.5). We then identify the extreme edges of that band and obtain a trunk whose base and top are regular pentagons.

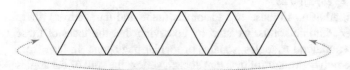

Fig. 2.5 Building a regular icosahedron, initial step

With the remaining 10 triangular faces, we build two pyramids with a pentagonal base and glue them to the trunk (Fig. 2.6). The result is a model of the icosahedron.

Step 2:   Construction of possible platonic solids with square faces.

We start by assuming that each vertex has three square faces. Thus, the angles of the faces incident on each vertex add up to $3 \times 90 = 270°$. From these hypotheses, we obtain the cube.

Then, with four square faces incident on a vertex, the sum of the angles would reach 360° and the polyhedron would not be convex. Therefore, there is only the cube as a regular polyhedron with square faces.

Fig. 2.6 Building a regular icosahedron, final step

Step 3:   Construction of possible platonic solids with pentagonal faces.

It is assumed that three pentagonal faces occur at each vertex. Since the internal angles of a regular pentagon measure 108°, then the angles of the faces incident to each vertex add up to $3 \times 108 = 324°$, and we get the dodecahedron (Fig. 2.7).

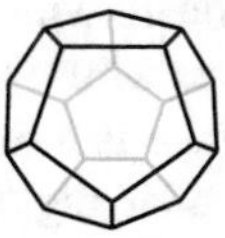

This exhausts the possible regular polyhedra with pentagonal faces, since with four faces incident on a vertex, the sum of the angles of the faces incident on each vertex would be greater than 360°.

Fig. 2.7 Dodecahedron

The relationship of Platonic solids (Fig. 2.8) with the fundamental elements, earth, fire, water and air is explained by Timaeus in a poetic way.

Fig. 2.8 The five platonic solids

According to Timaeus, the cube, due to its stability, represents the earth. The tetrahedron, the acutest body, represents fire. The icosahedron rolls like water and the octahedron represents air. The fifth element, the dodecahedron, with its 12 faces, one for each element of the zodiac, represents the cosmos.

Many of these Platonic beliefs had already been worshiped centuries before by the followers of Pythagoras. From the Pythagorean school come the oldest manifestations of understanding the cosmos through mathematics.

The Pythagoreans were very concerned with harmony and aesthetics. Note the elegance contained in the famous Pythagoras theorem (Fig. 2.9): the height of the right triangle, relative to the hypotenuse, defines two other right triangles, one to the left and one to the right of the height. All three triangles are similar in that they have the same angles. From this similarity comes a proof of the theorem.

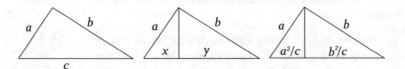

Fig. 2.9 Pythagoras theorem

If the legs have lengths a and b and the hypotenuse length c, with $c = x + y$, then $cx = a^2$ and $cy = b^2$. Indeed, from the similarity between the triangle on the left and the original triangle, we have $a/x = c/a$, that is, $cx = a^2$. Analogously, from the similarity of the triangle on the right to the original triangle, we have $b/y = c/b$, that is, $cy = b^2$. Therefore, the sum of the squares of the legs, $a^2 + b^2 = cx + cy = c(x + y) = c^2$, the square of the hypotenuse. Viewed in another way, the area of the square with the edge equal to the hypotenuse is the sum of the areas of the two squares whose edges are equal to the legs of the right triangle.

There is a more general form of the Pythagorean theorem that relates the areas of similar regions resting on the sides of a right triangle. In the right triangle of hypotenuse c and legs a and b, we say that regions resting on the three edges are similar if they differ only by a scalar factor; in other words, if there is a homothety between the regions. Let r_1 and r_2 be the homothety ratios between the edges of the right triangle; that is, $b = r_1 a$ and $c = r_2 a$. Since the triangle is a right triangle, we have $a^2 + b^2 = c^2$, that is, $a^2 + (r_1 a)^2 = (r_2 a)^2$. Therefore, $1 + r_1^2 = r_2^2$ (Fig. 2.10)

If the areas of the regions resting on the sides of lengths a, b and c of the right triangle are, respectively, A, B and C, then the dilations provide the following relations between the areas: $B = r_1^2 A$ and $C = r_2^2 A$. Hence, $A + B = (1 + r_1^2)A = r_2^2 A = C$.

Pythagoras Theorem Consider the right triangle of hypotenuse c and legs a and b. Consider similar regions, supported by the hypotenuse and the legs, whose areas are respectively C, A and B. Then, $C = A + B$.

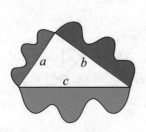

Fig. 2.10 Generalized Pythagoras

Pythagoras was fond of numbers and this made him notice numerical subtleties in both astronomy and music. For Pythagoreans, the numerology associated with pleasant sounds also occurred in the harmonious universe. Good music was appreciated as a major art and reproduced the harmony of a perfect universe. This fantastic numerology has been and still is studied by alchemists and esoterics. There is even a type of geometry, known as sacred geometry, that attracts many followers.

2.2 Arithmetic, Geometric and Harmonic Means

Timaeus' discourse on the perfection of the universe makes extensive use of certain numerology, in the Pythagorean style. Timaeus describes true rituals that mix faith with mathematics of certain ratios and proportions, chosen to describe the magnificent work of the creator, as they perceive coincidences, according to fervent believers, as incontestable. Three of these ratios between two quantities, say a and c, which appear in Plato's writings, are actually quite useful. They are the arithmetic mean b_1, the geometric mean b_2 and the harmonic mean b_3:

$$b_1 = \frac{a+c}{2} \qquad b_2 = \sqrt{ac} \qquad b_3 = \frac{2ac}{a+c}$$

The arithmetic mean b_1 is expressed as b_1 exceeds a by the same as c exceeds b_1; that is, $b_1 - a = c - b_1$. The geometric mean satisfies the ratio $a : b_2 = b_2 : c$. The harmonic mean b_3, which Timaeus uses when describing the composition of the soul, states that b_3 exceeds a with respect to a, just as c exceeds b_3 with respect to c. In symbols, $(b_3 - a) : a = (c - b_3) : c$.

One can also describe the harmonic mean b_3 of a and c as follows: the reciprocal of b_3 is the arithmetic mean of the reciprocals of a and c; that is, $1/b_3 = (1/a + 1/c)/2$.

A geometric construction of these three means is made in a semicircle of diameter $a + c$ (Fig. 2.11). The central vertical segment is the radius of the semicircle, so it has length $(a + c)/2$, which is the arithmetic mean of a and c. The right triangle with vertices A, B, C inscribed in the semicircle has height h and, as we know, $h^2 = ac$, that is, h is the geometric mean of a and c. The value of x in the diagram corresponds to the harmonic mean of a and c. To demonstrate this, firstly note that $x + y = (a + c)/2$.

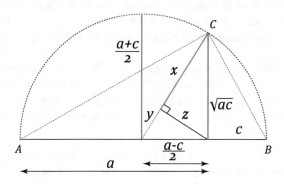

Fig. 2.11 Geometric construction of arithmetic, geometric and harmonic means

Now, applying the Pythagoras theorem to the right triangle of legs y and z and to that of legs x and z (Fig. 2.11), we obtain respectively $y^2 = (a - c)^2/4 - z^2$ and $z^2 = ac - x^2$. Hence, $y^2 = (a - c)^2/4 - (ac - x^2)$. From the equality,

$x + y = (a + c)/2$, we get $y^2 = (a + c)^2/4 - x(a + c) + x^2$. Equating these two expressions of y^2 yields $x = 2ac/(a + c)$, which is the harmonic mean of a and c.

These means are used in a variety of activities.

In Architecture
In his book entitled *I quattro libri dell'architettura,* published in 1570, the architect Andrea Palladio (1508–1580) describes some prototypes of rectangular rooms. When the height of the rectangular rooms corresponds to the means of the sides of the rectangle, arithmetic, geometric or harmonic, the most pleasant places would be obtained, according to Palladio.

In Music
Arithmetic and harmonic means appear in the construction of the Pythagorean musical scale.

A legend says that Pythagoras was inspired by hearing a blacksmith hit the anvil with different hammers, creating different sounds.

Stretching a string fixed at the ends, Pythagoras noted that the sound produced by making the string vibrate had a particular relationship with that produced by a string twice the size. Thus, the 2:1 ratio was designated an octave. Subdividing the vibrating string by half produces an octave higher, while doubling the length of the vibrating string one gets an octave lower. The octave was divided into two other parts, called the fifth and perfect fourth. The fifth was obtained by the arithmetic mean and the fourth by the harmonic mean between 2 and 1. The arithmetic mean is 3/2, and the harmonic is 4/3. Thus, the fifth is defined by the 3:2 ratio, while the fourth is 4:3. Therefore, 4:3:2:1.

Other Examples Using the Means

Example 1 A student leaves home for school at the same time. If walking at a constant speed of 2 km/h, she then arrives at 11 am. If walking at a constant speed of 6 km/h, she then arrives at 9 am. How fast must the student walk to arrive at 10 am? If the answer was the arithmetic mean of the speeds $(2 + 6)/2 = 4$, it is wrong. The correct answer is the harmonic mean of the two speeds 2 and 6, therefore 3 km/h.

Example 2 If a car travels a certain time interval T at a constant speed X, and travels through the same time interval T at a constant speed Y, then the average speed over the entire time $2T$ will be the arithmetic mean of X and Y. In fact, the distance traveled is the sum of $XT + YT$, so the average speed corresponds to this distance divided by $2T$.

If the car travels a distance L at a constant speed X and then travels the same distance L at a constant speed Y, then the average speed at which the car travels the distance $2L$ will be the harmonic mean between X and Y.

Indeed, the time it takes the car to do the first part of the course is L/X and the second part is L/Y. Therefore, the average speed will be the distance traveled, $2L$, divided by the time, $L/X + L/Y$; ie, $2L/(L/X + L/Y) = 2LXY/(LX + LY) = 2XY/(X + Y)$, so it is the harmonic mean of X and Y.

Example 3 (A Generalization of Example 1) If you leave home at time t_0, you arrive at school at time t_1 when walking at a constant speed x km/h, and arrive at time t_2, when walking at a constant speed y km/h. At which constant speed z km/h, as a function of x and y, should one walk to get to school on time $(t_1 + t_2)/2$?

The distance from home to school is equal to the speed multiplied by time, so $x(t_1 - t_0)$, which is equal to $y(t_2 - t_0)$. From this equality, it follows that the exit time $t_0 = (yt_2 - xt_1)/(y - x)$. But, $z((t_1 + t_2)/2 - t_0)$ is also the distance from home to school. Therefore, $z((t_1 + t_2)/2 - t_0) = x(t_1 - t_0)$. Substituting the value of t_0, we obtain $z = 2xy/(x + y)$, that is, z is the harmonic mean of x and y.

2.3 The Golden Proportion

Another famous ratio, the favorite of the mystics, is known as *the golden ratio* or *golden proportion*, and is defined as follows: a segment of length L is divided into two parts of lengths X (larger part) and $L - X$ (smaller part), we say that it is divided into the golden proportion if the whole L is for X as X is for $L - X$; that is, $L/X = X/(L - X)$.

In Book VI of Euclid Elements, the golden ratio is called *division into mean and extreme ratio*. Note that $X^2 + LX - L^2 = 0$, i.e., $(X/L)^2 + (X/L) - 1 = 0$, and therefore $X/L = (-1 + \sqrt{5})/2$.

Fig. 2.12 The Parthenon

The number $(-1 + \sqrt{5})/2$, approximately 0.618, is known as the golden number and is often denoted by the Greek letter ϕ, supposedly after the Greek architect Phidias (480 BC–430 BC). Indeed, Phidias is believed to have used the golden ratio in the project of the Parthenon in Athens (Fig. 2.12).

The reciprocal $1/\phi$ of the golden number is sometimes also known as the golden number and is denoted by the capital Greek letter Φ.

We have, $\Phi = 1 + \phi$. Indeed,

$$\frac{1}{\phi} = \frac{2}{-1 + \sqrt{5}} = \frac{2(1 + \sqrt{5})}{(-1 + \sqrt{5})(1 + \sqrt{5})} = \frac{1 + \sqrt{5}}{2} = 1 + \phi,$$

that is, Φ is approximately 1.618.

The rectangle of sides L and X, whichever values of L and X, such that $X^2 + LX - L^2 = 0$, is called the *golden rectangle*. The size of the rectangle does not matter, but rather the proportion of its sides.

Some people associate golden rectangles with a manifestation of the creator, because, according to this belief, many forms in nature exhibit this proportion. It is also said that it is the rectangle that most pleases our vision. On this subject, there is a very good article by George Markowsky [5].

In addition to these ancient beliefs, which even today have many followers, what is really beautiful about the golden ratio is the relationship that defines it: if a line segment is divided into two parts, a larger and a smaller one, then the division of the segment is golden when the entire length is to the larger part as the larger part is to the smaller.

Also attractive is the ruler-and-compass construction of a golden rectangle. This is done with a circle centered at the midpoint of one side of the square, as in Fig. 2.13.

There are many occasions in which the golden number appears and this ubiquity can explain the true adoration that some develop for the theme. Let us just list a few geometric curiosities of this proportion.

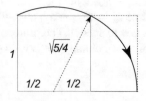

Fig. 2.13 Golden rectangle

2.3.1 Golden Rectangles and Architecture

Fig. 2.14 Le Corbusier and Einstein

Architect Charles-Édouard Jeanneret-Gris (1887–1965), better known as Le Corbusier, was an unconditional fan of the golden ratio. In his book *Le Modulor* [4], he defends that the architectural project should be guided by the golden number. Believing that the golden ratio manifests itself in various parts of the human body—for example, the navel would divide our height into the golden ratio—he claimed that built spaces should be shaped in the same proportion to harmonize with their inhabitants (Fig. 2.14).

Corbusier was so fascinated by his discovery that he sought out Albert Einstein at Princeton in 1946 to discuss his theory with the professor. In a passage in *Le Modulor,* Corbusier describes this encounter, which he says was not as fruitful as it could have been because he was nervous. In fact, out of emotion, he interrupted Einstein's speech and ended up not quite understanding whether, after all, Einstein had liked what he saw or not. Later, according to Le Corbusier, Einstein wrote a letter to him stating: *the system of proportions you presented made the bad difficult and the good easy*, whatever that means.

Figure 2.15 depicts Le Corbusier's *Unité d'habitation* in Marseille (a) and Berlin (b).

(a)   **(b)**

Fig. 2.15 Le Corbusier unité d'habitation

2.3.2   The Golden Number in the Regular Pentagon

In any regular pentagon, the diagonals intersect at points that divide them in golden
proportion. In fact, the isosceles triangles of vertices ABC and DBC (Fig. 2.16a)
have the same interior angles, so they are similar. The first has base length L and
sides X, while the second triangle has base X and sides $L - X$. From the similarity,
the proportion follows: $L/X = X/(L - X)$. Hence, $X^2 + LX - L^2 = 0$ proves the
claim.

An interesting model of a regular pentagon with the diagonals crossing can be
made with a knotted rectangular ribbon of paper (Fig. 2.16b).

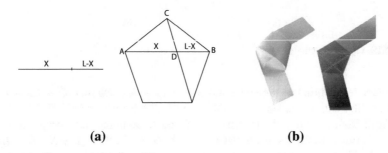

(a)   **(b)**

Fig. 2.16 Golden proportion

The composition of the five diagonals of the regular pentagon is called the
pentagram, a symbol that some find very mystical. A pentagram with one of the
vertices pointing upwards (Fig. 2.17a) is the symbol that identifies many sects,
including the Pythagorean school in ancient Greece, and the meaning of *good* is
attributed to it. The pentagram with a downward-facing apex and an inscribed goat's
head is used as a symbol of *evil*, and identifies sects associated with black magic.
The drawing in Fig. 2.17b was originally published in 1897 in the book *La Clef de
la Magie Noire* by occultist Stanislas de Guaita (1861–1897).

Fig. 2.17 The pentagram

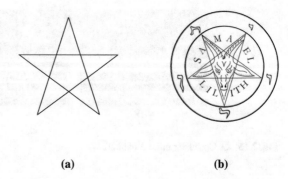

(a)   (b)

2.3.3   Golden Rectangles and the Sum of Squares

Given a golden rectangle, we attach a square to the longest side. The resulting rectangle is also a golden rectangle.

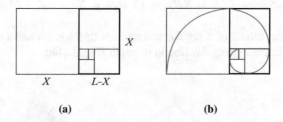

(a)   (b)

Fig. 2.18 Golden rectangles

In fact, if the initial rectangle has sides X and $L - X$ (such that $X^2 + LX - L^2 = 0$), then after adding the square of side X we will have a rectangle of sides L and X (Fig. 2.18a). Note that the lengths $L + X$ and L are in the same proportion as L and X; in other words, knowing that $L/X = X/(L - X)$, then $(L + X)/L = L/X$. Similarly, if we remove a square whose side is equal to the smaller side of a golden rectangle, the remaining rectangle will also be golden.

In the composition of golden rectangles obtained by the coupling of squares, by drawing an arc of a quarter of a circumference in each of the squares, a curve similar to a spiral is obtained (Fig. 2.18b).

Those most enchanted with golden rectangles see this spiral shape in many phenomena, from the shell of a nautilus to the shape of our galaxy.

2.3.4 The Golden Number in Trigonometry

In trigonometry, the golden number helps to calculate sines and cosines of multiples of the angle $\pi/5$.

In fact, we learned at school the sine and cosine values of the so-called special arcs, namely, $\pi/2$, $\pi/3$, $\pi/4$ and $\pi/6$, as well as the null angle. These angles appear on the faces of some Platonic solids when cut in half; $\pi/2$, $\pi/3$ and $\pi/6$ are the interior angles of half of an equilateral triangle, while $\pi/2$ and $\pi/4$ of the triangle obtained as half of a square. What about $\pi/5$?

As can be seen in the table of values of cosines and sines (Fig. 2.19), there is no place for $\pi/5$ without breaking the beautiful symmetry!

In addition to the aesthetic motif, the value of $\cos(\pi/5)$ is an irrational number and, unlike the special arcs, its square is also irrational.

Let us prove that $\cos(\pi/5) = \Phi/2$.

θ	$\dfrac{\pi}{2}$	$\dfrac{\pi}{3}$	$\dfrac{\pi}{4}$	$\dfrac{\pi}{6}$	0
$2\cos(\theta)$	$\sqrt{0}$	$\sqrt{1}$	$\sqrt{2}$	$\sqrt{3}$	$\sqrt{4}$
$2\sin(\theta)$	$\sqrt{4}$	$\sqrt{3}$	$\sqrt{2}$	$\sqrt{1}$	$\sqrt{0}$

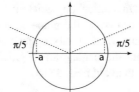

Fig. 2.19 Sine and cosine of special angles

Let $a = \cos(\pi/5) > 0$ then $\cos(4\pi/5) = -a$ (Fig. 2.19). Let $b = \cos(2\pi/5)$. So, $b = \cos(\pi/5 + \pi/5) = \cos^2(\pi/5) - \sin^2(\pi/5) = 2\cos^2(\pi/5) - 1$. Therefore, $b = 2a^2 - 1$.

Also, $-a = \cos(4\pi/5) = \cos(2\pi/5 + 2\pi/5) = 2\cos^2(2\pi/5) - 1$. So, $-a = 2b^2 - 1$. Therefore, $b + a = (2a^2 - 1) - (2b^2 - 1) = 2(a + b)(a - b)$. Hence, $a - b = 1/2$. So, $a - 1/2 = b = 2a^2 - 1$; that is, $4a^2 - 2a - 1 = 0$.

Finally, we obtain the value of $\cos(\pi/5)$ as the positive root of the quadratic equation $4a^2 - 2a - 1 = 0$; that is, $a = (1 + \sqrt{5})/4 = \Phi/2$.

Note that the angle $3\pi/5$ occurs at each vertex of a regular pentagon, and $\pi/5$ is the trisection provided by the two diagonals that fall on each vertex. Therefore, angle $\pi/5$, the one excluded from the set of special arcs, has a close relationship with the golden number Φ.

2.3.5 The Golden Number and the Quasicrystals

Filling a plane with small pieces is an ancient art known as mosaic.

The oldest mosaics were built in the 3rd millennium BC in Mesopotamia, but it was in the Byzantine Empire, between the 11th and 15th centuries, that the art flourished.

In addition to being decorative, mosaics are also important records. The mosaic map of Jerusalem, found on the floor of St.George's Church in Madaba, Jordan, dates from the sixth century and describes details of the buildings at that time in that city (Fig. 2.20).

When the pieces of a mosaic are polygons and cover an entire plane, the mosaic is called tiling. If the patterns of a tile are repeated, it is called periodic tiling. There are tiles that are never repeated and that is why they are called aperiodic. In this case, a region of the tiling can be translated in the plane and will never coincide with another region of that tiled plane.

Fig. 2.20 Mosaic map of Jerusalem

Mathematical physicist Sir Roger Penrose created several aperiodic tiles. One of them can cover the plane with only two types of pieces, formed by two isosceles triangles with unitary edges and bases ϕ and Φ, the golden numbers obtained in making the pentagram (Fig. 2.21). There is a rule to arrange these prototypes in order to avoid periodicity.

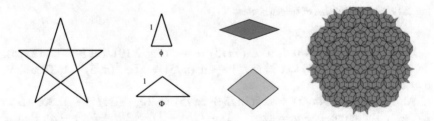

Fig. 2.21 Penrose tiling

In 1981, the British crystallographer Alan Lindsay Mackay used a Penrose tiling in two and three dimensions to predict a new kind of ordered structure not allowed by traditional crystallography and it was called quasicrystals.

Until 1984, Penrose's aperiodic tiles served as mathematical objects to describe quasicrystals; that is, polyhedral structures that, unlike crystals, are never repeated in translation. It was materials engineer Dan Shechtman who first communicated the incredible existence of a quasicrystal. The famous biochemist Linus Pauling (1901–1994), Nobel Prize winner in Chemistry in 1954, reacted rudely to Shechtman's discovery, claiming that there are no quasicrystals, only quasicientists. For several years, until his death, Pauling continued to criticize that work severely, leading to a certain boycott of Dan Shechtman. However, in 2011, Shechtman finally had his discovery recognized by the chemists' community, and the alleged quasicientist was also awarded the Nobel Prize in chemistry.

2.3.6 The Golden Number and the Means

Let A, G and H, respectively, be the arithmetic, geometric and harmonic means of two quantities a and c. That is,

$$A = \frac{a+c}{2} \qquad G = \sqrt{ac} \qquad H = \frac{2ac}{a+c}$$

The numbers A, G and H verify the equality $G^2 = AH$ and the inequalities $H \leq G \leq A$ can be observed in the diagram of the geometric construction of the means (see Fig. 2.11).

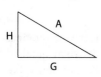

Fig. 2.22 Right triangle

According to Angelo Di Domenico [3], the means A, G and H are the hypotenuse and legs of a right triangle if, and only if, $A/H = \Phi$. Indeed (Fig. 2.22), from Pythagoras theorem, it follows that the triangle of sides A, G, H is a right triangle if, and only if, it verifies $A^2 = H^2 + G^2$. Dividing by H^2 both sides of the equality and using $G^2 = AH$, we obtain $(A/H)^2 = 1 + (G/H)^2 = 1 + (AH/H^2)$. Hence, $(A/H)^2 = 1 + (A/H)$, that is, $A/H = \Phi$.

2.3.7 Golden Rectangles and the Regular Dodecahedron

Fig. 2.23 Golden rectangles and a regular dodecahedron

If three identical golden rectangles are arranged in a two-by-two orthogonal position, the three will have a common point, as shown in Fig. 2.23. Properly joining the 12 vertices of this composition with edges of the same length, we obtain a regular icosahedron, the Platonic symbol of water. The 12 vertices of these same three rectangles two by two perpendicular will also be the centers of the 12 faces of a regular dodecahedron, the symbol of the cosmos. Thus, the regular dodecahedron has a very close relationship with the golden ratio, in addition to its pentagonal faces.

2.3.8 The Golden Number and the Fibonacci Sequence

The sequence named after Leonardo Fibonacci (1170–1250) is the numerical sequence $1, 1, 2, 3, 5, 8 \cdots$ whose nth term is the sum of the two previous terms. Like the golden ratio, the Fibonacci sequence attracts many mystics. They believe that various natural phenomena, from plant growth to rabbit reproduction, are described by this numerical sequence.

There is indeed an interesting relationship between the golden number and the Fibonacci sequence. If F_n denotes the nth term of the Fibonacci sequence, then $F_n = F_{n-1} + F_{n-2}$. Dividing this equality by F_{n-1}, we obtain:

$$\frac{F_n}{F_{n-1}} = 1 + \frac{F_{n-2}}{F_{n-1}}$$

When n is very large, in other words, when n tends to infinity, if the left quotient of the equality F_n/F_{n-1} tends to a non-zero number F, then the right quotient F_{n-2}/F_{n-1} will tend to the number $1/F$. Hence, when n approaches infinity, the above equality becomes $F = 1 + 1/F$, that is, $F^2 = F + 1$. Therefore, $F = \Phi$; in other words, the quotient of adjacent terms of the Fibonacci sequence will tend to the golden number when n tends to infinity.

Italian artist Mario Merz (1925–2003) is the author of works based on the Fibonacci sequence. In the Igloos series, Merz collects hemispheres whose radius follow the numerical sequence [6].

2.3.9 Johannes Kepler

Our universe is as beautiful as it is complex, so it is no wonder brilliant minds like Plato have dedicated their lives to the quest for cosmic order, and perhaps for an explanation of our origin and destiny, beyond this brief period of life on planet Earth. Perhaps that is why in this quest, which continues to this day, logic sometimes gives way to faith. In certain dialogues, Plato is even fanciful. For example, in *The Symposium*, also translated as *The Banquet*, at the time of Aristophanes' speech, he describes how in the early days we were spherical for aesthetic reasons; nothing more symmetrical and beautiful than a sphere, he said. We rolled around. The spheres were of three sexes—male, female and hermaphrodite. Until one day, due to the insolence of the primordial human beings, Zeus cut them in half, generating a pair of men from the masculine sphere, a pair of women from the feminine, and a man and woman from the hermaphrodite sphere. The halves then went out into the world. When one half meets its other half, the passion is immediate; the halves complete each other and want to be together forever.

Plato's work created a veritable army of followers who profoundly influenced the development of Western civilization. One of these, the mathematician and astronomer Johannes Kepler (1571–1630), made no secret of his adoration for Platonic solids. Kepler was passionate about the idea of a universe perfectly designed by a God geometer, and this enormous passion stayed with him until the end of his life.

As a fervent Christian that he was, Kepler believed that his works should be dedicated to understanding the divine work of creating the universe. Letters he sent to his colleagues took the form of current scientific papers. Some of these are important contributions to understanding the solar system, such as the well-known three laws of planetary motion published in 1609, as well as an incredibly accurate description of the orbit of Mars. However, other articles emerged as a result of the delusions of a radical believer.

Kepler's famous article, entitled *Mysterium Cosmographicum,* published in 1597, presents a model of the solar system with the six planets that were known at the time. It comprises the five concentric Platonic solids, in a chosen order, interspersed with spheres circumscribed to each of the polyhedra and with a sixth sphere inscribed in the first polyhedron of the sequence.

The article contains a drawing of the model (Fig. 2.24) by Christoph Leibfried (1566–1635). Kepler calculated the ratios of the radii of those spheres and related them to the orbits of the six planets in the solar system.

Fig. 2.24 Christoph Leibfried 1597

In the preface, Kepler wrote: *I undertake to prove that God, in creating the universe and regulating the order of the cosmos, had in view the five regular bodies of geometry as known since the days of Pythagoras and Plato, and that he has fixed according to those dimensions, the number of heavens, their proportions, and the relations of their movements* [7, p. 114].

Unfortunately, the data from Mysterium Cosmographicum did not correspond to those obtained experimentally by the Danish Tycho Brahe (1546–1601) and neither to the Polish Nicolas Copernicus (1473–1543). After the discovery of other planets in the solar system, Kepler's model was rejected by the scientific community. According to Kline: *His predilection for fitting the universe into a preconceived mathematical pattern, however, led him to spend years in following up false trails* [7, p. 113].

Kepler also searched polyhedra for mystical properties. An example is a polyhedron composed of two regular tetrahedrons, which Kepler named *stella octangula.* Mystics call it *merkaba* and believe that the object has fantastic properties and represents a vehicle to transport body and soul to other dimensions.

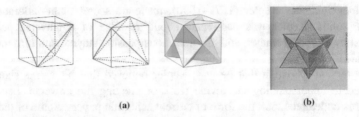

(a)   (b)

Fig. 2.25 (**a**) Stella octangula   (**b**) Leonardo da Vinci

To build the stella octangula, or merkaba if you prefer, you start with a cube, and on each of the six square faces you choose one of the two diagonals.

By choosing the diagonals appropriately, they will shape the six edges of a regular tetrahedron (Fig. 2.25a). By superimposing the two tetrahedrons obtained from the two possible choices of diagonals, a stella octangula is obtained.

A drawing of the stella octangula (Fig. 2.25b) by Leonardo da Vinci (1452–1519) was published in the book *De divina proportione* written by the Franciscan friar Luca Pacioli (1445–1517) in 1509.

Alchemists such as Isaac Newton and religionists like Johannes Kepler found in geometry a vehicle to rationalize their beliefs. Thus, it cannot be excluded that part of the development of geometry is related to this search for a divine order in the cosmos.

References

1. Atyah, M. and Sutcliffe, P; *Polyhedra in Physics, Chemistry and Geometry,* Milan Journal of Mathematics 71, 33–58 (2003).
2. Euclid; *The thirteen books of Euclid's elements*, translated by Sir Thomas Heath, Cambridge University Press 1908.
3. Di Domenico, Angelo; *The golden ratio, the right triangle and the arithmetic, geometric and harmonic means*, Math. Gazette 89, p. 261 (2005).
4. Le Corbusier; *Le modulor*, Harvard University Press, 2nd edition 1958.
5. Markowsky, George; *Misconceptions about the golden ratio*, The College Math. Journal 23, 2–19 (1992).
6. Merz, Mario; https://www.maxxi.art/arte/
7. Kline, Morris; *Mathematics in Western Culture*, Oxford University Press 1953.
8. Plato; *Timaeus*, translated by Benjamin Jowett (1871) http://classics.mit.edu/Plato/timaeus.html

Chapter 3
Geometry: From Disorder to Order

3.1 Euclidean

In prehistory, 3500 BC, a set of techniques was already used to calculate areas, volumes, measure edges of regions and solve problems involving these concepts. That is what is written on cuneiform clay tablets found in southern Mesopotamia, which is Iraq today.

Fig. 3.1 (**a**) Euclid's Elements 1570 (**b**) Papyrus fragment 100AD (**c**) Principia 1687

In these records, created by the magnificent Sumerian civilization, the triumph of order over chaos was celebrated. These Sumerian techniques, later improved in Babylon and ancient Egypt, were incorporated by the Greek civilization into a deductive system, which was later called Geometry.

Thales (624 BC–558 BC), Pythagoras (570 BC–495 BC), Plato (427 BC–347 BC), Euclid (325 BC–265 BC), Archimedes (287 BC–212 BC) are some names of genius Greeks who left us an extraordinary intellectual heritage.

The deductive method consists of obtaining theorems from basic assertions and logical deductions. It had profound implications in Western civilization development. Several areas of knowledge were remarkably influenced, especially the

T. Marar, *A Ludic Journey into Geometric Topology*,
https://doi.org/10.1007/978-3-031-07442-4_3

mechanics of Isaac Newton (1643–1727), his Principia Mathematica (Fig. 3.1c), and the philosophy of Baruch de Spinoza (1632–1677).

Around 300 BC, Euclid organized a masterful work called The Elements, in which he records the mathematical knowledge of the time [3]. Euclid's Elements contain the so-called axiomatic Euclidean geometry. After the Bible, it is considered the most translated, published and studied book in the Western world.

The story that is told about the paths that this work took to reach our textbooks is full of adventures. For centuries, generations of wealthy individuals invested in paying for copies of Euclid's work, always made by specialized scribes, which led to its survival, since the papyrus on which it was written deteriorated in a few years. A fragment found in Egypt (Fig. 3.1b) by English archaeologists in the late nineteenth century is the closest to the original and dates back to the year 100 of the Christian era. The first Latin edition is from the year 1300 and the first English edition (Fig. 3.1a) from 1570, both translated from Arabic.

Consisting of 13 books, Euclid's work is very original, but far from perfect.

Euclid begins Book I with a list of definitions, some of which are somewhat obscure, and only by reading the text can one get an idea of the author's intentions. The first definitions of Book I are:

Definition 1 *A point is that which has no parts.*

Definition 2 *A line is a breadthless length.*

Definition 3 *The extremities of a line are points.*

Definition 4 *A straight line is a line which lies evenly with the points on itself.*

A length with no width suggests the one-dimensional character of the lines, just as the point is zero-dimensional. However, as the German scholar, Christoph Friedrich von Pfleiderer (1736–1821) pointed out, someone who happens not to know what a straight line is would hardly benefit from Definition 4 [3, p. 168].

Historians are still debating why Euclid would have elaborated such a confusing definition. Some wonder why he did not do like Archimedes, who considered the straight line to be the shortest line between its endpoints? One possible explanation lies in the differences between the motivations of the two philosopher-scientists. In fact, today Archimedes would qualify as a mathematical physicist, while Euclid would be a pure mathematician.

Next Euclid defines two-dimensional objects.

Definition 5 *A surface is that which has length and breadth only.*

Definition 6 *The extremities of surfaces are lines.*

Definition 7 *A plane surface is a surface which lies evenly with straight lines on itself.*

Again, the definitions suggest that a surface is
a two-dimensional object and finite in size. The
plane surface is as confusedly defined as its one-
dimensional counterpart.

Other definitions, among the 23 in Book I, are
notable for their lack of clarity.

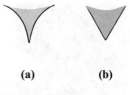

(a)   **(b)**

Fig. 3.2 Angles

Definition 8 *A plane angle is the inclination to one
another of two lines in a plane which meet one another
and do not lie in a straight line.*

Definition 9 *When the lines containing the angle are straight, the angle is called
rectilineal.*

Today, what we call angle, as Euclid in most of his book, is the rectilineal angle
(Fig. 3.2b), if we understand the region between the lines as inclination. But what is
the purpose of definition 8? Was Euclid implying the existence of a non-Euclidean
geometry? We will come back to this topic shortly.

Many problems in the physical world are understood and solved using Euclidean
geometry and mathematical models. The passage from abstract concepts, such
as Euclid's definitions, to mathematical models of the physical universe requires
many compromises. Generations of philosophers have discussed this topic and the
debate continues to this day. Far from being a trivial matter, it is from the search
for mathematical models that applications of geometry in the solution of natural
problems arise.

Everything that exists in the physical world is three-dimensional. In fact, all we
see and touch is the outside, the two-dimensional surface of solid objects in nature.
Despite this, some three-dimensional objects serve well as physical models of zero,
one, or two-dimensional geometric objects. As we have already mentioned, a sheet
of paper is a good model of a plane surface, as it is long, wide and of negligible
thickness. When we consider point A on the surface, we mark it on the sheet of
paper with a pen. That little bit of ink represents that which has no parts. We do
the same for another point B. From point A to point B, there are infinite lines
contained on the plane surface, in particular a straight line. To represent this straight
line between A and B on the sheet of paper, we draw it with the pen, highlighting
all the points of the straight line between A and B. We use a ruler, which helps
in tracing what will be a representation of a segment of a straight line in the plane
modeled by the sheet of paper. A ruler can be made out of a piece of wood, checking
its straightness observing it with your eye, thus assuming that the light paths are
straight lines. So, there are many agreements to obtain representations of abstract
geometric objects.

Fig. 3.3 Bertrand Russell

In an interview, the British polymath Nobel laureate Bertrand Russell (1872–1970) (Fig. 3.3) said that his first contact with Euclid's work was at age 11, having his older brother as a tutor. For Russell, this would have been one of the great events of his life, as exciting as his first love. *I had not imagined there was anything so delicious in the world,* confesses Russell, for whom until the age of 38 mathematics would have been his main source of happiness. However, in 1902, Russell published a three-page article in which he listed severe criticisms of Euclid's lack of clarity [7]. A true love-hate relationship.

Perhaps inspired by the same criticisms, Russell gives us a funny description of pure mathematics: *...thus mathematics may be defined as the subject in which we never know what we are talking about, nor whether what we are saying is true* [5].

In an attempt to explain how a theory as abstract as geometry can be useful in solving physical problems, Russell refers to two categories: one is the pure geometry of Euclid and the other is geometry as a branch of physics. In the first type, consequences are deduced from the axioms, without asking whether the axioms are true. Everything is obtained by logic and without the need for pictures, although they help a lot in the construction of the discourse. The second type of geometry is an empirical science in which axioms are derived as an inference from measurements and differ from those of Euclid.

According to Russell, geometry as a branch of physics facilitates the understanding of the applications of geometry in problem solving. However, we should add that the laws of physics do not extend to geometric objects: the same object can occupy two positions in space, just as two objects can occupy the same position.

In addition to mathematical models as a way of using geometry in solving problems, we can totally ignore the nature of the objects involved in a given problem and focus only on how the objects are related. These relationships are described as postulates or axioms.

Imagine two sets of objects of a completely different nature. Assuming that the relations between the objects of one of the sets are the same as those between the objects of the other set, then whatever conclusions it is possible to deduce for the objects of the first set, the same is deduced for the second.

The following two examples show how to use axiomatic geometry to deduce theorems from some postulates that describe how objects are related, regardless of the nature of the objects. In the first example, the objects belong to the world of ideas, while we can say that the second example refers to our real world.

Example 1 (According to David Hilbert [4])

Postulate 1: A straight line contains at least two points.
Postulate 2: For every two points A, B, there is a straight line that contains each of the points A, B.

Postulate 3: For any three non-collinear points, there is exactly one plane that contains them.

Postulate 4: If two points of a straight line lie in a plane, then every point of the straight line lies in the plane.

Theorem *For a straight line and a point outside it, there is exactly one plane that contains both.*

Proof From Postulate 1, there are two points on the straight line. These two points and the one outside the line are three non-collinear points. From Postulate 3, there is a plane determined by these three points. This plane contains two points of the straight line determined by them (Postulate 2) so, from Postulate 4, the plane contains the entire straight line.

Example 2

Postulate 1: A contractor finances at least two corrupt deputies.

Postulate 2: For any two corrupt deputies, there is exactly one contractor that finances them.

Postulate 3: For any three corrupt deputies not financed by the same contractor, there is exactly one public enterprise that finances them.

Postulate 4: If two corrupt deputies are financed by a public enterprise, then the contractor who finances them is financed by the public enterprise.

If we accept the above postulates, then:

Theorem *For a contractor and a corrupt deputy not financed by that contractor, there is exactly one public enterprise that finances both.*

The protagonists of the first example, points, lines and planes are replaced, in the second example, by corrupt deputies, contractors and public enterprises, respectively. The membership relation in example 1 is replaced by the financing operation in example 2. The theorem deduced in the first example is rewritten in the second example. Once the first theorem has been proved, there is no need to prove the second, due to the analogy between the two examples. What matters in the deductive method is the way in which objects are related, even if they are of a different nature.

In geometry, some propositions can be heavily based on geometric constructions. To illustrate this, we will show three beautiful theorems attributed to three great geometers, namely, Thales, Euclid and Archimedes.

1. Tales

Proposition 20 from Euclid's Book III is known as the central and inscribed angle theorem.

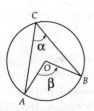

Consider a circle of radius r and center O. Let A, B and C be three points on the circle. The quadrilateral of edges OA, AC, CB and BO defines a central angle AOB, call it β, and an inscribed angle ACB, call it α (Fig. 3.4). The theorem states that, for any circle and points A, B, C on it, $\beta = 2\alpha$. Even if β is greater than $180°$, and hence some edges of the quadrilateral are smaller than the radius of the circle.

Fig. 3.4 Tales

When the three points A, O and B are aligned, then $\beta = 180°$, and the result is known as Thales' theorem. In this case, the quadrilateral becomes a triangle inscribed in the semicircle, with $\alpha = 90°$, therefore a right triangle. In other words, any triangle inscribed in a semicircle is a right triangle.

A proof of the proposition, that is, a sequence of arguments that convinces a non-believer of the validity of the result $\beta = 2\alpha$ makes use of other results. One such result is known as *Pons asinorum*. In Latin, Pons asinorum means bridge of asses. The nickname of this proposition, which is the fifth in Book I of Euclid, is said to have come from the Middle Ages. As it is the first significant result of *The Elements*, it is believed that anyone who did not understand the proof should not continue reading. It is also known as the base angle theorem: *In isosceles triangles the angles at the base are equal to one another;* that is, for a triangle that has two sides of the same length, then two of its three angles are also equal.

The following proof of the Pons asinorum is attributed to Pappus of Alexandria (290–350). His argument consists of considering the two isosceles triangles ABC and BAC, as reflected triangles (Fig. 3.5a). Since these two triangles have equal sides, then all three angles are respectively equal. In particular, the angle of vertex A of triangle ABC and vertex B of triangle BAC are equal. Therefore, the base angles of any isosceles triangle are equal.

The mathematician Lewis Carroll considered this argument of Papus as a joke, an *Irish bull*, since the same triangle has to be simultaneous in two different places. In response to this objection, Bertrand Russell would say: *So what!*

To prove the central and inscribed angle theorem, we note that in the quadrilateral $OABC$, the edges OA and OB have lengths equal to radius r.

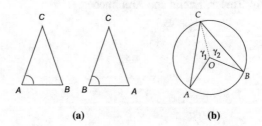

(a)   (b)

Fig. 3.5 Pons asinorum

Consider a new edge OC (Fig. 3.5b). It also has length r. So, the quadrilateral $OABC$ is divided into two isosceles triangles, angle α is divided into two, say, α_1 and α_2. Also, angle $\gamma = 2\pi - \beta$ is divided into γ_1 and γ_2. Since the triangle AOC is isosceles, then the angles of the base are equal and, therefore, the angle at vertex A is equal to α_1. The same happens with the isosceles triangle BOC, in which the angle at vertex B is equal to α_2.

Knowing that the sum of the interior angles of any triangle is equal to π, we obtain the following equalities: $\gamma_1 + 2\alpha_1 = \pi$ and $\gamma_2 + 2\alpha_2 = \pi$. Therefore, $\beta = 2\pi - \gamma = 2\pi - (\gamma_1 + \gamma_2) = 2\pi - ((\pi - 2\alpha_1) + (\pi - 2\alpha_2)) = 2(\alpha_1 + \alpha_2) = 2\alpha$.

As a consequence of this theorem, we have that the sum of the opposite angles α and β of any quadrilateral inscribed on a circle is equal to π. In fact, twice these angles are central angles that add up to 2π (Fig. 3.6).

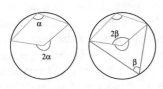

Fig. 3.6 Opposite angles

2. Euclid

Consider a circle with center O and two chords (segments inside the circle, whose extremities lie on the circle) crossing at a point C inside the circle (Fig. 3.7a). Euclid's theorem states $a_1 a_2 = b_1 b_2$.

The proof makes use of the previous theorem.

The two triangles $CA_1 B_1$ and $CA_2 B_2$, formed by the

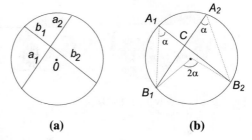

(a)   **(b)**

Fig. 3.7 Euclid

chords and the dotted segments (Fig. 3.7b), are congruent triangles because all their respective angles are equal. Therefore, $a_1/b_1 = b_2/a_2$.

3. Archimedes

Arbelos is the name given to the region delimited by three tangent semicircles, two inside the third, and whose centers are collinear.

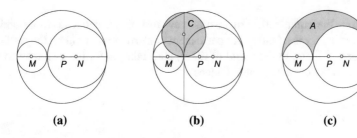

(a)   **(b)**   **(c)**

Fig. 3.8 Archimedes arbelos

Consider the arbelos created by semi circles with centers M, N and P and whose area is equal to A (Fig. 3.8c). Consider the segment perpendicular to the diameter of the larger circle, passing through the point of tangency of the two inner circles, and whose ends are on the larger circle (Fig. 3.8b). Let C be the area of the circle that has half of that perpendicular segment as its diameter, which we call circle C. Archimedes' theorem states that the areas of the arbelos and the circle C coincide; that is, $A = C$.

For a proof, consider r_1 and r_2 the radii of the inner circles, r the radius of the largest circle and let s be the radius of circle C. Thus, $r = r_1 + r_2$. From Euclid's theorem applied to the following chords given by the diameter of the largest circle and the perpendicular segment, we obtain: $(2s)(2s) = (2r_1)(2r_2)$. Hence, $s^2 = r_1 r_2$. Thus, $C = \pi s^2 = \pi r_1 r_2$. Subtracting the areas of the smaller semicircles from the area of the larger semicircle, we obtain area A of the arbelos: $A = (\pi r^2)/2 - \pi(r_1^2 + r_2^2)/2$. Finally, substituting $r^2 = r_1^2 + r_2^2 + 2r_1 r_2$ yields $A = \pi r_1 r_2 = C$.

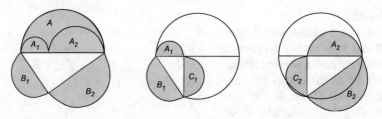

Fig. 3.9 Roger B. Nelsen arbelos

Another proof of this result, based only on figures and the generalized Pythagorean theorem (Chap. 2), was published by Roger B. Nelsen [6].

The generalized Pythagoras' theorem will be applied to three right triangles, whose regions supported on the legs and hypotenuses are semicircles (Fig. 3.9). The semicircles on the legs have areas A_1, A_2 and C_1, C_2 and on the hypotenuses the semicircles have areas B_1, B_2.

Note that the choice gives $C_1 + C_2 = C$.

Thus, $A + A_1 + A_2 = B_1 + B_2$, $B_1 = A_1 + C_1$ and $B_2 = A_2 + C_2$. Therefore, $A = C_1 + C_2 = C$.

And so, from the combination of results, new results are deduced, in an endless sequence. Geometry is a field under permanent construction, or rather, the results of geometry are all there, in the world of ideas, all it takes is for the geometer to find them, as it was believed in ancient Greece.

3.2 Euclidean and Non-Euclidean

Having attracted heavy criticism, Euclid's work was rewritten many times.

In 1899, David Hilbert (1862–1943) (Fig. 3.10) published a version of his *Grundlagen der Geometrie* [4], correcting several inaccuracies in Euclid's book. While Euclid listed five axioms, Hilbert demanded twenty (we use four of them in Example 1, above). It was not long before Hilbert also found inaccuracies in his own work, of which he published other versions, to be finally satisfied with the seventh.

From the five axioms of Euclid, the fifth became known as the postulate of parallels and was, for a long time, a reason for dispute among mathematicians due to its propositional character. There were many unsuccessful attempts to deduce the fifth axiom from the other four. In 1868, the Italian mathematician Eugenio Beltrami (1835–1900) demonstrated the independence of the fifth axiom from the other four, ending the controversy.

Fig. 3.10 David Hilbert

Hilbert kept Euclid's fifth postulate in his list of 20 axioms, but wrote it as follows: *In a plane α there can be drawn through any point A, lying outside of a straight line r, one and only one straight line which does not intersect the line r. This straight line is called the parallel to r through the given point A.* This is known as Playfair axiom named after the Scottish mathematician John Playfair (1748–1819).

Axiomatic systems that fulfill the conditions of non-contradiction and independence allow for the construction of theories. Even negating some axioms of a system of independent axioms without contradictions can give rise to coherent theories.

Particularly, in the case of Euclidean geometry, negations of the parallel postulate allowed the creation of perfectly coherent and distinct geometries from that of Euclid, much to the despair of the defenders of the oneness of truth.

One of the first to register alternatives to the fifth axiom was the Persian Omar Khayyam (1048–1131) (Fig. 3.11). He produced texts, long ignored, which today are considered notable contributions to the study of geometry.

Fig. 3.11 Omar Khayyam

Most of our history books feature Omar Khayyam as a Muslim poet, writer of the Rubaiyats. The Portuguese poet Fernando Pessoa (1888–1935), under the heteronym Ricardo dos Reis, signed several poems based on Khayyam's quartets. However, in addition to being a poet, Khayyam was a great mathematician.

It seems that Khayyam, annoyed by the apparently inconsequential attitude of Euclid, who had been concerned with demonstrating obviousness leaving other statements without a satisfactory demonstration, decided to write his own work *On difficulties of postulates and definitions of Euclid,* in the year 1077. In it Khayyam includes five new postulates, in place of the parallel postulate. Based on this he demonstrates propositions 29 and 30 from Book I of Euclid, which are the first in which Euclid uses the fifth postulate. Furthermore, Khayyam demonstrates propositions of a new geometry, other than the Euclidean one.

In another of Khayyam's works, one finds what is now known as the Blaise Pascal (1623–1662) triangle and also a geometric solution to certain algebraic equations of the third degree, four centuries before the appearance of a general solution by Niccolò Tartaglia (1499–1557) and Gerolamo Cardano (1501–1576). Definitely, Omar Khayyam's mathematics was way ahead of his time.

Fig. 3.12 Saccheri 1733

The jesuit Gerolamo Saccheri (1667–1733) wrote a book that, once accepted by the Inquisition and later by the Society of Jesus, was published two months after his death (Fig. 3.12). In it, Saccheri presents studies that also contributed to the development of geometry, repeating much of Khayyam's writings.

It is not known whether Saccheri had access to Khayyam's work. Saccheri negates the fifth postulate and comes very close to formulating a non-Euclidean geometry.

Several mathematicians participated in the formalization of non-Euclidean geometries, especially Nikolai Lobachevsky (1792–1856), János Bolyai (1802–1860) and Bernhard Riemann (1826–1866).

These new geometries are consistent theories just as the Euclidean geometry, which was unique for over 2000 years. This turning point can be seen as the beginning of *modernity* in geometry.

Hyperbolic geometry and elliptic geometry (also called spherical geometry) are two of such non-Euclidean geometries, obtained as a result of the negation of the fifth axiom: through a point outside of a given line passes a line parallel to the given one. There are two ways to negate this: through a point outside a given line, more than one line passes parallel to the given one, or less than one, hence none. Having an appropriate interpretation of the concept of a straight line, the first negative gives rise to hyperbolic geometry, in which through a point outside a *straight line* pass infinitely many parallels to the given line, and from the second comes elliptic geometry, in which, for a point outside a *straight line*, there is no parallel to the given line.

The planes of these non-Euclidean geometries, namely the hyperbolic plane and the elliptical plane (Fig. 3.13), when represented in our three-dimensional space, take on the appearance of curved surfaces: in the hyperbolic case, a hyperbolic paraboloid, which is a surface of negative curvature (in relation to the Euclidean plane, which has zero curvature) and in the elliptical case, a sphere, which is a surface of positive curvature and all the correspondent *straight lines* are arcs circles of maximum diameter, also called *geodesics*.

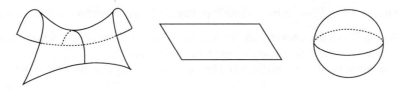

Fig. 3.13 Hyperbolic, Euclidean and elliptical planes

Thus, through two points on the sphere, there is a geodesic that corresponds to the straight line defined by two points on the Euclidean plane (Fig. 3.14).

In general, the two points on the sphere together with its center form three non-aligned points through which a single plane passes. This plane intersects the sphere in a circle of maximum diameter and two arcs of this circle join the two points considered. The geodesic is the smaller of the two arcs.

In aerial navigation, geodesics help to define the shortest trajectories between two points on our planet.

Fig. 3.14
Geodesic

Through the north and south poles of the sphere, there are infinite geodesics, as they are aligned with the center of the sphere.

Since the geodesic arc in the sphere corresponds to the segment of a straight line in the plane, then in the spherical geometry any two *straight lines* intersect; there are no parallels. Also in spherical geometry Euclid's axiom 2 is not valid, which states that the lines extend as much as you like. Other constraints occur in spherical geometry. For example, there is not necessarily an isosceles triangle for a given base, as is the case in Euclidean geometry.

Non-Euclidean geometries can look strange because we use Euclidean geometry as the underlying space for our representations. The opposite could also seem strange. For example, if we consider our planet represented by a sphere, then the most direct air traffic routes are geodesic arcs between airports. However, on the world map, a representation of spherical geometry in the Euclidean plane, the same routes are not projected into straight lines, which in the plane are the shortest paths between pairs of points.

There are other representations of the planes of non-Euclidean geometries in Euclidean space. One of them, which represents the lines of the hyperbolic plane in the Euclidean plane, is known as the Poincaré Disk (Fig. 3.15). This model was explored in several drawings by the Dutch illustrator Maurits Escher (1898–1972) [2].

Fig. 3.15 Poincaré disk

Let us continue reading the definitions of Book I of Euclid, interpreting some of them in non-Euclidean geometries.

In The Elements, definitions 13 and 14 read as follows:

Definition 13 *A boundary is that which is an extremity of anything.*

Definition 14 *A figure is that which is contained by any boundary or boundaries.*

Again, Euclid does not explain his definitions very well. In this case, the meaning of extremity is not clear, as well as the term to be contained.

Definition number 19, on the other hand, deals with a particular type of figure:

Definition 19 *Rectilineal figures are those which are contained by straight lines, trilateral figures being those contained by three, quadrilateral those contained by four, and multilateral those contained by more than four straight lines.*

Some authors believe that this definition was not found in Euclid's original, as he always refers to figures by the number of angles and not the number of edges. In both hyperbolic and elliptical geometry, objects corresponding to Euclidean straight lines appear as curved when plotted on the respective planes, which in turn are represented in three-dimensional Euclidean space.

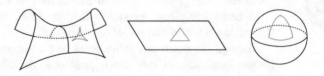

Fig. 3.16 Triangles in hyperbolic, Euclidean and elliptical planes

Thus, using Euclid's definition 8 of angle (not necessarily rectilineal), it turns out that triangles in the hyperbolic plane have the exotic property of the sum of its internal angles being less than 180°, while in elliptical geometry, the sum of the internal angles of any triangle is greater than 180° (Fig. 3.16).

The name of the great German mathematician Carl Friedrich Gauss (1777–1855) is mentioned by several historians in the formulation of the new geometries, but his participation is controversial. Mathematical historian Jeremy Gray argues that Gauss was more interested in the geometry of physical space than in the underlying axiomatics. According to Gray, on the topic of non-Euclidean geometries, Gauss presented himself more as a scientist than a mathematician.

One of the most discussed issues in this subject concerns whether or not Gauss had made empirical tests to find that the sum of the angles of an enormous triangle defined by the peaks of three known mountains in Germany added up to less than 180°. This would imply that the geometry of the universe would be hyperbolic. However, even in a huge triangle, the deviation that the sum of the interior angles would have from 180° would be tiny, and almost always imperceptible to any measuring instrument.

In his 1921 article *Geometrie und Erfahrung,* Einstein describes Bernhard Riemann's proposals for a hyperspheric universe with an elliptical geometry.

Finally, it is worth mentioning that the terms ellipse and and hyperbola were coined in ancient Greece by the geometer Apollonius of Perga (262 BC–190 BC). Today, as figures of speech, ellipse and hyperbola refer respectively to the omission and excess of terms in a sentence. In the case of non-Euclidean geometries, a circle of radius r in the hyperbolic plane has a perimeter greater than $2\pi r$, while in the elliptic geometry plane, the perimeter of the circle is less than $2\pi r$ (Fig. 3.17).

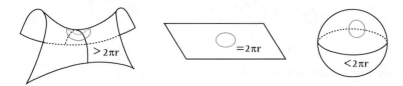

Fig. 3.17 Circles in hyperbolic, Euclidean and elliptical planes

3.3 Felix Klein

Fig. 3.18 Felix Klein

Since the development of mathematics is essentially evolutionary, moments of paradigm shifts are rare.

In the process of formalizing geometries, there was a notable revolution; it is a result of the work of Felix Christian Klein (1849–1925) (Fig. 3.18) presented at the chair examination in 1872 at the University of Erlangen. His thesis became known as the *Erlangen Programm*. For Klein, a geometry in a space is the result of the action of an equivalence relation (congruence) defined by a set of suitable transformations. A geometry organizes the objects of space into classes; all objects of the same class being equivalent according to the equivalence relation that defines that geometry.

The maxim of Klein's program: geometries without axioms. This is an example of what Gaston Bachelard (1884–1962) defined as the new scientific mind [1].

When applying a geometry to a given space, something analogous to the mythological descriptions of the creation of the universe occurs when a God geometer promotes the passage from chaos to order. In fact, in a space in which a geometry is installed, objects are organized according to the equivalence relation that defines the geometry. It is the passage from disorder (Fig. 3.19a) to order (Fig. 3.19b).

More precisely, according to Klein, a geometry is constituted by a set of objects (points) S and a group G of transformations of S. A figure is a subset of S. A g-property is a property invariant by the action of group G. Figures P and Q are g-congruent if there is a transformation g of G that takes P to Q. Since g-properties are invariant by the action of G, it follows that if P and Q are g-congruent, then they have the same g-properties.

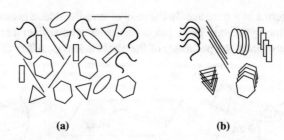

(a)   (b)

Fig. 3.19 From disorder to order

Thus, there are several geometries that can be considered in the same space, each one of them establishing an organizational process.

The postmodernity in geometry begins with Klein's program. Projective geometry, hyperbolic geometry and topology are some of these non-Euclidean geometries easily described by the Erlangen Programm.

Depending on the transformations that define the geometry, our perception of the objects in the space changes. For example, in hyperbolic geometry, through a point outside a straight line, infinite lines pass parallel to the given line.

From Klein's point of view, Euclidean geometry is the result of the equivalence relation defined by isometric transformations (translation, rotation and reflection): two objects are congruent if one can be moved over the other and they coincide.

Because Euclidean geometry is the study of space whose objects have properties that do not change when rigid motion is applied to them, then this geometry establishes an organizational process in which the equivalence relation between the objects maintains its metric characteristics, such as lengths, areas, angles, etc.

Fig. 3.20 Euclidean order

Once the Euclidean order is applied to a space, its objects are organized into classes defined by isometries; that is, each object represents all those equal to it by isometric transformations (Fig. 3.20).

In Proposition 4 of Book I of Euclid's Elements, known as the congruence side, angle, side of triangles, there is a process of superposition of figures. Although the concept of congruence is not explicit in Euclid's masterly work, one has the impression that Euclid had anticipated Klein's proposals 2000 years earlier.

Moving a body requires having a place for it to occupy. The notion of place, space and matter is discussed in one of the few surviving writings by the Greek philosopher Archytas (428 BC–347 BC). According to Archytas, a body occupies a place and cannot exist without it. For Archytas, space is where all phenomena occur, and the notion of place is primordial.

As we have seen, the use of geometry in understanding physical space, in solving concrete problems, requires a process of associating ideas with things. This association is quite subtle, as things in the physical world are crude when compared to ideas, which are perfect. When Euclid describes his fundamental concepts, such as: a point, that which has no parts; a line, that which has only length; and a surface, that which only has length and width, it shows the zero-dimensional character of the point, one-dimensional line and two-dimensional surface. Point, line and surface are concepts and do not exist as objects in nature. However, we can represent them through certain three-dimensional physical models—as are, in fact, all the objects around us—that have characteristics similar to those described in the concepts.

Fig. 3.21 Tripartite Unity

For example, in the field of materials science, two-dimensional materials are referred to, when it means extremely thin materials.

In the arts, the metal sheet that Max Bill (1908–1994) used to make the Tripartite Unity sculpture (Fig. 3.21) has such a reduced thickness compared to its width and length that it serves as a representation of a surface in Euclid's sense. This work was awarded first prize at the first *Bienal de Artes de São Paulo* in 1951 and it belongs to the Museum of Contemporary Art of the University of São Paulo—USP. In the next chapter, we will analyze the surface represented by this sculpture, shedding light on the title that the artist gave to it.

Claims that curved lines and twisted surfaces are non-Euclidean are frequent. Such objects can belong to Euclidean as well as non-Euclidean geometry, since what will determine one thing or another is the relational properties between the objects, defined by the geometry of the underlying space.

Fig. 3.22 Heydar Alyev Center

The building of the Heydar Aliyev Center in Azerbaijan (Fig. 3.22), by architect Zaha Hadid (1950–2016), has a roof that resembles a portion of the model of the hyperbolic plane represented in our physical space, which does not mean that the work has a hyperbolic geometry.

Just as Euclidean geometry is defined, according to Klein, by rigid movements, non-Euclidean geometries are defined by other equivalence relations, distinct from isometries. Topology, for example, is a geometry obtained by the equivalence relation defined by *homeomorphisms*, which consist of continuous transformations that can be continuously undone. In topology, the representation of objects is made with an imaginary material that is perfectly deformable.

Fig. 3.23 Topological order

Thus, all open lines are homeomorphic to the straight line and all closed lines are homeomorphic to the circle. In other words, one-dimensional objects are topologically organized into only two classes: closed lines, represented by the circle, and open lines, represented by the straight line (Fig. 3.23).

In the two-dimensional case, the list of closed surfaces, that is, surfaces finite in size and without boundary, has existed since 1920. The list is divided into two, namely, the orientable surfaces (see Chap. 4), which are those that have a well-defined interior and exterior, and non-orientable surfaces, which are those that contain a Möbius strip (see Chap. 6), which is a famous surface named after the German mathematician August Möbius (1790–1868).

Felix Klein was an extraordinary mathematician and the boldness of his proposals, including the Erlangen Programm, prevented them from being universally accepted by his contemporaries. Today Klein is honored with a non-orientable surface that bears his name: the Klein bottle (Fig. 3.24a). This surface is shaped like a bottle and it can be obtained from two Möbius strips.

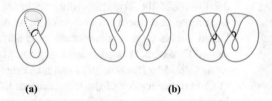

(a)   (b)

Fig. 3.24 (a) Klein bottle (b) identifying two Möbius strip along their boundaries

A model of the strip is constructed with a long rectangle whose vertical edges are identified after a half-turn (180°) rotation. The Möbius strip has only a single curve as its boundary. A three-dimensional space model of the Klein bottle is obtained by identifying two Möbius strips along their boundaries (Fig. 3.24b).

3.4 Points at Infinity

The concept of space is the subject of extensive philosophical discussions. The difficulty of defining space intrinsically has led to some philosophers assuming the concept as an *a priori,* a presupposition. However, creating geometries in the Kleinian way can help to perceive a space, which changes depending on the geometry considered in that space. On the one hand, according to Klein, congruence relations define a geometric order among objects in different spaces; on the other hand, geometry defines the properties of space, making the pair space+geometry an entity whose understanding is more straightforward.

The space with its geometry will be called *the underlying space* of that geometry. When we represent such spaces in the Euclidean environment, we are able to compare them. For example, hyperbolic geometry has as its underlying plane a surface of negative curvature, when compared to the Euclidean plane which has zero curvature.

Another geometry, created from visual observation, called projective geometry, has an underlying space somewhat different from that of Euclidean geometry. In projective geometry, points at infinity, also called improper points, are considered. A point at infinity is the direction of a straight line. Two parallel lines have the same point at infinity, so they meet there.

Fig. 3.25 Point at infinity

Projective geometry is believed to have been developed by Renaissance painters who realized that, in the world we see, parallel lines suggest a meeting point. When we are facing a long railway, the tracks, which must be good representatives of parallel lines so that the train does not derail, seem to meet at a very distant point (Fig. 3.25). This is how the world is projected onto our retina: the parallels meet at their improper point, the direction common to the two parallel lines.

A Euclidean straight line together with its improper point is called a *projective line*. A circle serves as a representation of the projective line. In fact, a Euclidean line has a direction, its point at infinity. This point is reached either by traversing the straight line in one way or in the opposite way. A Euclidean line can be represented by a straight line segment without ends, called an open segment. There is a one-to-one correspondence between the points on a Euclidean line and the points on an open segment; in other words, there are as many points on the open segment as on the entire straight line. The improper point of the straight line is represented by the extremities of the open segment, and they must be identified (Fig. 3.26a) in the projective geometry.

(a)   (b)

Fig. 3.26 (a) Projective line   (b) Projective plane

To build a model of the plane of the projective geometry, which we call the *projective plane,* we must add to the Euclidean plane all its points at infinity; that is, the directions of all its straight lines. For this, consider a model of a Euclidean plane represented by a disk without its circular boundary, called an open disk. There is a one-to-one correspondence between the points on an open disk and the points on the Euclidean plane. The circular boundary points will represent the points at infinity. In fact, a straight line passing through the center of the disk intersects the circular boundary at two points, which represent the same point at infinity. Thus, in the projective plane model, these two diametrically opposite points, also called antipodal points, must be identified. All lines parallel to the one through the center of the disk will have the same point at infinity. Therefore, the points of the circular boundary of a disk are all the points at infinity of the Euclidean plane. Therefore, the projective plane is the Euclidean plane represented by an open disk, together with all the identified pairs of antipodal points on the circular boundary (Fig. 3.26b). Models of the projective plane in three-dimensional Euclidean space are presented in Chap. 6.

Projective geometry, in which parallels meet (at infinity), in addition to agreeing with the geometry seen in the physical world, creates an organizational process that simplifies certain studies. An example of this simplification can be seen in the study of plane curves.

Fig. 3.27 Ellipse, parabola and hyperbola

After the straight lines, the most commonly used plane curves in geometry are the conics: ellipses, parabolas and hyperbolas (Fig. 3.27). While straight lines are described by first-degree algebraic equations, also called linear equations, conics are second-degree algebraic curves. For example, in a Cartesian coordinate system with origin O and axes Ox and Oy, the equation $x + 2y - 1 = 0$ describes the set of coordinate pairs (x, y) that are the points of a straight line; the equation $x^2 + 2y^2 - 1 = 0$ describes the set of coordinate pairs (x, y) that are the points of an ellipse.

Conics are given this name because they are obtained from the intersection of a straight circular cone with a plane (Fig. 3.28a). Ellipses, parabolas and hyperbolas are the classic conics.

In the Euclidean plane, they configure very different curves. Associated with a parabola is a line called the symmetry axis; hyperbola, on the other hand, has a close relationship with two concurrent lines, called asymptote lines. Hyperbolas are made out of two pieces, called branches, in the Euclidean plane.

However, in the projective plane, ellipses, parabolas, and hyperbolas are all closed curves. In fact, the ellipse has no points at infinity, it is completely contained in the Euclidean plane. The parabola has one improper point, given by the direction of its axis of symmetry. Going over a parabola, to return to the starting point we must go through its point at infinity (Fig. 3.29b). Hyperbola has two improper points, which are the two directions of its asymptote lines. Going over a hyperbola, to return to the starting point we must go through its two points at infinity.

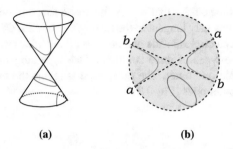

(a)   **(b)**

Fig. 3.28 **(a)** Conic sections   **(b)** Ellipse, parabola and hyperbola in the projective plane

Suppose you start a route over the hyperbola at a point on its upper branch (Fig. 3.29c). We followed this branch until we reached the first improper point, the direction of one of the asymptote lines. We reappear at the opposite end of that asymptote line. Continuing the course, we will go over the lower branch until we reach the other improper point, the direction of the other asymptote straight line. We reappear at the opposite end of this asymptote line and finally return to the starting point on the upper branch.

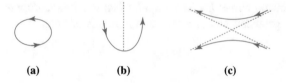

(a)   **(b)**   **(c)**

Fig. 3.29 Elliptical, parabolic and hyperbolic trajectories

The earliest known records of conics were written by Apollonius of Perga, who some believe was a student of Euclid.

His treatise on conics is considered one of the most profound works of antiquity.

The name of Apollonius is also related to a classic problem called the Apollonius' problem: to find a circle that passes through three given objects, the three objects must belong to the set consisting of points, straight lines and circles. While the meaning of the term passing through is obvious in the case of points, passing through a line here means being tangent to the straight line and the same goes for the circle.

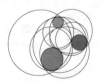

Fig. 3.30 Appolonius circles

In the simplest case, Apollonius' problem is to find a circle that passes through three (non-aligned) points. In the most interesting case, three disjoint circles are given, whose centers are not aligned, and a new circle that is simultaneously tangent to the three given circles must be found (Fig. 3.30).

For each set of three circles that satisfy the above assumptions, there are exactly eight circles simultaneously tangent to the three given circles. Of the eight solutions, one has all three given circles inside and one has all outside. The other six solutions are divided into two types of three solutions. One type tangents the three circles, having two inner and one outer. The other type tangents the three circles, having one inner and two outer ones (Fig. 3.31).

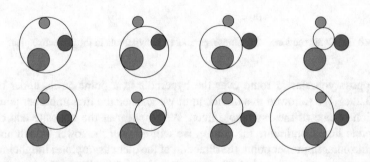

Fig. 3.31 Solutions to Appolonius problem

In 1676, Newton wrote: *If I have seen further, it is by standing on the shoulders of giants.* The geometer and astronomer Apollonius of Perga was certainly one of those giants.

References

1. Bachelard, Gaston; *La formation de l'esprit scientifique*, J. VRIN 1934.
2. Escher, M.C.; https://mcescher.com/gallery/mathematical/
3. Euclid; *The thirteen books of Euclid's Elements*, translated by Sir Thomas Heath, Cambridge University Press 1908.
4. Hilbert, David; *The Foundations of Geometry*, translated by Townsend, E., The Open Court Publ. Co. 1902.
5. Knowles, Elisabeth ed.; *Oxford dictionary of modern quotations*, Oxford University Press 2007.
6. Nelsen, Roger B.; *Proof without Words: The Area of an Arbelos*, Mathematics Magazine 75, p. 144 (2002).
7. Russell, Bertrand ; *The Teaching of Euclid,* The Mathematical Gazette 33, 165–167 (1902).

Chapter 4
Topology

4.1 A Kind of Geometry

As we saw in the previous chapter, Felix Klein's program describes a geometry as a system that organizes the objects of a given space. Through an equivalence relation, objects are arranged into congruence classes. All objects of the same class are equivalent, so any object of a given class serves as a representative for all objects of that class.

In topology, the congruence classes are defined by *homeomorphisms;* loosely speaking, they are continuous transformations that can be continuously undone. Two objects are topologically equivalent if one can be continuously deformed into the other, and vice versa. Thus, contrary to the rigidity of objects in Euclidean geometry, in topology the objects are modeled by an imaginary material which is perfectly deformable, and therefore allows for any continuous deformation. Questions in topology are of qualitative nature rather than quantitative ones. Consequently, in topology there is no mention of lengths, areas, angles, etc. If we stretch or shrink an object, its topological characteristics do not change. Furthermore, other operations with the models of the objects, like cutting and gluing (see Sect. 4.1.2 below), are allowed and they are important when the ambient space containing the objects are taken into consideration.

If, on the one hand, homeomorphisms do not maintain metric properties of the objects, they maintain something invariant, which, as we shall see, is important in the classification of objects according to their shape.

Although this definition of topological congruence is quite opaque, like Euclid's definitions, it will be clarified with many examples.

From the topology point of view, objects that we usually take as different are identified, and this creates a curious geometric perception. For example, a triangle and a circle are topologically congruent; in fact, any polygon can be continuously deformed into a circle. Therefore, any polygon is homeomorphic to a circle. Likewise, any open line is homeomorphic to a straight line.

© The Author(s), under exclusive license to Springer Nature Switzerland AG 2022   47
T. Marar, *A Ludic Journey into Geometric Topology*,
https://doi.org/10.1007/978-3-031-07442-4_4

Note that if we remove a point from an open line, it breaks into two parts. The same does not happen with a closed line, which remains connected after one of its points is removed. This shows that open and closed lines are topologically distinct.

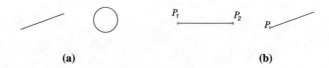

(a)   **(b)**

Fig. 4.1 One dimensional topological objects

Thus, the topological classification of one-dimensional objects is very simple: there are only two classes of one-dimensional objects; namely, open lines and closed lines. Objects of the first class can be represented by the straight line, and those of the second class by the circle. We are referring to one-dimensional objects with no boundary. Boundaries of lines are the endpoints, two points, or just one (Fig. 4.1b).

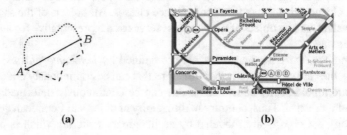

(a)   **(b)**

Fig. 4.2 Connecting points

In topology, the study of one-dimensional objects differs enormously from that of Euclidean geometry. In fact, while in Euclidean geometry questions about one-dimensional objects are of various natures, such as length, curvature, torsion, angles, areas, in topology they boil down to questions about open or closed lines. For example, are two points connected or not?

In topology, the connection between two points, A and B, can be made with a straight line or, equivalently, with any other open line (Fig. 4.2a).

The distance between points A and B is irrelevant in topology. A consequence of this topological property was ingeniously applied in the design of the London underground maps by Harry Beck (1903–1974) in 1933.

His idea is still reproduced today by most subways in the world (Fig. 4.2b). Prioritizing the connection over the distance between two stations makes a topological map a more functional representation than a geographic map. In fact, for the people who take the subway, it is essential to know the connections between the stations.

Surfaces, two-dimensional objects, are also classified topologically. A surface is called a closed surface if it has no boundary and is limited; that is, finite in size. The list of closed surfaces has been around since 1920 and each congruence class defines what we will call the object's shape.

A classic example in topology is the continuous deformation that transforms the surface of a donut into the surface of a one handle mug. Donut and one handle mug have the same shape. This is why a topology specialist can be defined as a person who can bite a mug at teatime thinking that it was a donut.

Closed three-dimensional objects can be classified topologically, but the list has not yet been completed.

4.1.1 A Brief History

The beginning of topology is attributed to the problem of the seven bridges of Königsberg (Fig. 4.3).

In this Prussian city, there were seven bridges over the Pregolya River that allowed the crossing of two islands (two of the seven original bridges were destroyed during a bombing in August 1944). The possibility of walking around the city, crossing all seven bridges only once, was questioned. The mathematician Leonhard Euler (1707–1783) showed the impossibility of such a route by appealing to a topological diagram and its congruences.

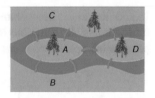

Fig. 4.3 Seven bridges

Let us follow Euler's reasoning for solving this problem and enjoy the beginnings of a new way of thinking in mathematics.

Initially, Euler observes that the choice of path within each region is not relevant to the problem. Then, he creates an abstract model in which each region is represented by a point (vertex) and the bridges connecting the regions are represented by segments (edges) that connect the points.

The resulting diagram is now called a *graph* and is widely used in several areas of mathematics (Fig. 4.4) . Euler synthesizes the problem by subtracting any irrelevant information using his graphs.

Fig. 4.4 Euler graph

When entering a region through a bridge, a hiker must leave through another bridge. Therefore, a graph that represents a path along which each Königsberg bridge is crossed only once, must have an even number of incident edges at each vertex, one entering the region and the other exiting; except for the initial and final vertex of the trajectory. If the begin and end of the path coincide, then to satisfy the Königsberg bridge problem, no vertex should have an odd number of incident edges.

Thus, the ingenious Euler solves the problem of the seven bridges of Königsberg (without leaving his house), demonstrating its impossibility, since in his diagram representing the islands and bridges, the vertices have an odd number of incident edges. Using that diagram Euler created a geometry in which the distance between two points does not matter, but whether or not they are connected. This is considered the beginning of topology, which for a long time was called *analysis situs*.

We can consider variations of the seven bridges problem. For example, if we eliminate the bridge connecting regions B and D from the diagram (see Fig 4.4), then yes, we can walk across all six bridges just once. For example, follow the path $C \rightarrow A \rightarrow B \rightarrow A \rightarrow C \rightarrow D \rightarrow A$, leaving C and arriving at A. Moreover, if we add another bridge connecting regions B and D; that is, add an eighth bridge to the original seven, then the problem will have a solution. There will be a route that goes through all eight bridges, starting at A and finishing at C, crossing each bridge only once.

4.1.2   Cutting and Gluing

In addition to the homeomorphisms that act as continuous deformations, there are other admissible operations that facilitate the topological classification. One of these permissible operations is *cutting* the object followed by *gluing*.

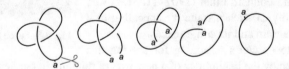

Fig. 4.5 From a trefoil knot to a circle

For example, a knotted line in three-dimensional space is topologically equivalent to a circle. Although it is not possible to deform a knot continuously in three-dimensional space to transform it into a circle, we can cut the knot at a point, creating two points, unknot it, and finally reconnect the two points created in the cut, forming a circle (Fig. 4.5).

The mere existence of a knotted curve (one dimensional object) is associated with three-dimensional space. The study of knotted curves in three-dimensional space is called Knot Theory. Many results of this theory are applied in other areas, as distinct as astronomy and microbiology.

The same knot could be undone without the need for cutting and gluing, if the curve is in four-dimensional space (see Chap. 5). Although representation in three-dimensional space is comfortable for us, in topology, objects exist intrinsically, without necessarily being located in an ambient space.

Cutting and gluing is also useful when classifying higher dimensional objects. While a cut on a line (one-dimensional object) gives rise to two points (zero-dimensional objects), on a surface (two-dimensional object) a cut will give rise to two edges (one-dimensional objects).

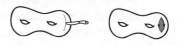

Fig. 4.6 Surface cutting

The pairs of edges, generated by cuts in surfaces, must be oriented according to the direction of the cut (Fig. 4.6). This orientation must be respected when proceeding with the gluing, so as not to change the topological type of the surface.

4.1.3 Basic Surfaces

In the topological study of surfaces, the cylinder and the Möbius strip play a key role. In fact, from these two surfaces and some simple operations, all the topological congruence classes of closed surfaces are obtained. Both the cylinder and Möbius strip (also known as Möbius band) are obtained from a rectangular figure (thus, a rectangular piece of Euclidean plane, or rather, a perfectly deformable film), identifying a pair of opposite edges, directly (cylinder) or after a 180° turn (Möbius strip). The boundary of a cylinder consists of two circles while the boundary of the Möbius strip is a closed line (therefore, topologically a single circle) (Fig. 4.7).

Fig. 4.7 Cylinder and Möbius strip

The cylinder and the Möbius strip are not topologically equivalent, there is no homeomorphism that transforms one into the other, so they belong to distinct topological congruence classes. Although the cylinder and Möbius strip models are both made of the same material (a rectangular figure), the action in the gluing is different.

Topology, this spectacular geometry, detects the different gestures in the making of models!

The set of surfaces is divided into two types; namely, orientable surfaces and non-orientable surfaces. An orientable surface has two sides, like the cylinder, which can be colored with two colors; one color for the inside and the other for the outside. Only one color is necessary for coloring a Möbius strip; it is a one-sided surface. The non-orientable surfaces are the ones that contain a Möbius strip.

We will build models of any closed surface, in three-dimensional space, using the cylinder, the Möbius strip and certain gluing techniques.

For example, by pasting the circular boundaries of two disks on the two circular boundaries of a cylinder, one obtains (topologically) a sphere (Fig. 4.8a), whose symbol is S^2.

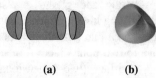

Fig. 4.8 Sphere and projective plane

The Möbius strip has only one circle as a boundary. By pasting the circular boundary of a disk along the boundary of the Möbius strip, a non-orientable surface called the *projective plane* (Fig. 4.8b), whose symbol is P^2, is obtained. This is not an easy operation to understand. In this case, a line of self-intersection appears. In fact, an extra dimension is needed to perform this gluing without creating self-intersections. In Chap. 5, which deals with the fourth dimension, details of this operation will be presented.

When we can built a model of a closed surface without self-intersections then we say that the surface *embeds* in three-dimensional space. The precise definition of an embedding of a surface in a given space is very technical, and it is not necessary for the moment.

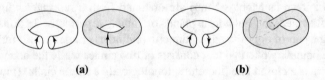

(a)   **(b)**

Fig. 4.9 (a) Torus   (b) Klein bottle

Another surface that derives from the cylinder is obtained by identifying the points on one of the circular boundaries with the points on the other one. Taking into account the orientation of each boundary circle, we obtain the surface called the *torus* (Fig. 4.9a), whose symbol is T^2. The torus T^2, the sphere S^2 and all closed orientable surface embed in three-dimensional space.

If, when identifying the boundary circles of the cylinder, we invert the orientation of one of the circles, we obtain the non-orientable surface called the *Klein bottle* (Fig. 4.9b), whose symbol is K^2. The Klein bottle can also be obtained by identifying two Möbius strips along their circular boundaries, point by point (see Fig. 3.24 in the previous chapter).

As it happens in the representation of the projective plane, also for the Klein bottle, and indeed for any closed non-orientable surface, their models in three-dimensional space always present self-intersection; that is, closed non-orientable surfaces does not embed in three-dimensional space (the Möbius strip is not closed, it has boundary).

Therefore, from cylinders and Möbius strips, and through gluing operations, we obtain the closed surfaces: sphere S^2, torus T^2, projective plane P^2 and Klein bottle K^2. The sphere and the torus are orientable, hence defining two sides; namely, an interior and an exterior in three-dimensional space. The projective plane and the

Klein bottle are non-orientable; they contain a Möbius strip. Some authors use the term one-sided surface when referring to non-orientable surfaces. In Chap. 5, it is explained why the term one-sided surface is meaningless for closed non-orientable surfaces.

Next, we will define another topological operation with surfaces and using the surfaces already obtained, we will describe all closed surfaces topologically.

4.1.4 Connected Sum of Surfaces

Connected sum is an operation involving two surfaces, F_1 and F_2, and the result is a third surface F_3. The surface F_3 is denoted by $F_1\#F_2$ and obtained by removing a disk from each, F_1 and F_2, and then gluing the two surfaces together along the boundaries left by removing the disks.

In topology, removing a disk from a surface is the same as removing a single point. In fact, after removing a point from a surface, whose model is made with a perfectly deformable material, we can enlarge the hole and create a boundary, as if we had removed a disk.

For example, the Klein bottle is the connected sum of two projective planes. In fact, as we have seen, the projective plane is obtained from the Möbius strip by pasting a disk along the circular boundary. Thus, removing a disk from the projective plane, a Möbius strip is obtained. The two strips identified along their boundaries yields the Klein bottle. Therefore, $K^2 = P^2\#P^2$ (Fig. 3.24 in the previous chapter).

The sphere is a neutral element of this sum; that is, $F\#S^2 = F$. In fact, by removing a disk from a sphere, the result is also a disk.

The connected sum of two tori $T^2\#T^2$ is a surface called a bitorus (Fig. 4.10).

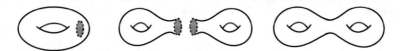

Fig. 4.10 Connected sum

4.1.5 The Fundamental Identity

The connected sum operation verifies the following equality:

$$T^2\#P^2 = P^2\#P^2\#P^2$$

This equality is called *the fundamental identity of the topological classification of closed surfaces*. It will be useful when we list the set of all closed surfaces. Note that in the operation of a connected sum of surfaces, unlike the sum of numbers, cancellation does not apply; e.g., canceling one P^2 on each side of the fundamental identity results in an absurd.

Let us show a geometric proof of the fundamental identity.

Initially, we construct the connected sum of T^2 with P^2 minus one point. Removing a point is necessary to facilitate the representation of the projective plane. In fact, by removing a point from P^2, one obtains a Möbius strip. As a connected sum requires the removal of a disk from each surface involved, let us construct a particular model of the torus with one disk removed, which will be crucial in proving the fundamental identity.

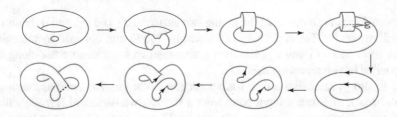

Fig. 4.11 Francis torus

In the sequence of figures (Fig. 4.11), we start with a torus, made of a perfectly deformable material, from which a disk has been removed. We expand the hole left by the disk removal until only a small ring remains. We cut the ring and deform it until we reach a flat ring with the two oriented cut edges, one edge on each boundary of the flat ring. Then, these boundaries are deformed to identify the edges produced by the cut. The result is a topological representation of a torus from which a disk has been removed. This representation of the torus with a disk removed can be found in George Francis' beautiful book *A topological picturebook* [3]. We will call it a Francis torus.

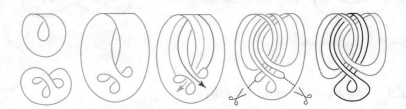

Fig. 4.12 Connected sum of a Möbius strip and a Francis torus

The connected sum of the torus with a projective plane is obtained from the fusion of a Francis torus with a Möbius strip.

Applying an adequate deformation to the surface $T^2 \# P^2$ minus one point, we arrive at a surface that, after two cuts, generates three Möbius strips (Fig. 4.12). This concludes the geometric proof of the fundamental identity.

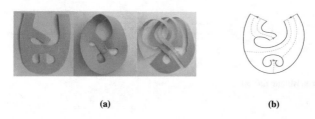

(a)   (b)

Fig. 4.13 Paper model of the connected sum of a Möbius strip and a Francis torus

Professor and magician José Luis Rodríguez from the University of Almeria-Spain [6] produced a paper model of this last operation (Fig. 4.13a). The reader can reproduce the model and, after the two identifications, cuts along the dotted lines (Fig. 4.13b) will produce the three Möbius strips that make up the connected sum of the torus with the projective plane.

4.1.6 Planar Models

The cylinder and the Möbius strip were represented by rectangular figures with pairs of opposing edges identified. This process of flattening a surface and representing it by a polygonal figure (in this case a rectangle and its interior) identifying pairs of edges, is called a planar model of the surface.

Given a closed surface; that is, a surface finite in size and without boundary, it can be flattened after a finite number of cuts. What guarantees this is a theorem that transcends the scope of this text.

We must imagine the surface modeled by a perfectly deformable film and the minimal number of cuts for flattening it will depend, say, on the culinary skills of the topologist.

Each cut along a line creates two edges (Fig. 4.14). So, in the end, the planar model will be a polygonal figure with an even number of edges. The

Fig. 4.14 Surface cutting

direction of the cut must be registered for future gluing in the surface reconstruction. Let us look at some examples.

Example 1 Torus planar model

When cut properly, the torus is turned into a cylinder (Fig. 4.15a). Then, a cut along the length of the cylinder yields a planar model (Fig. 4.15b).

Thus, the torus is flattened with two cuts. Each cut corresponds to two oriented edges. Thus, the planar model of the torus is a rectangular figure.

The torus can be reconstructed using its planar model. Indeed, if the opposite edges of the rectangle with two arrowheads (Fig. 4.15b) are identified, it generates a

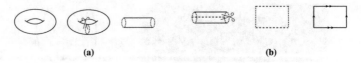

(a)   **(b)**

Fig. 4.15 Torus planar model

cylinder. Then, identifying the circular boundaries with one arrowhead, we end up
with a torus.

In the reconstruction, points of
opposite edges are identified. For
example, the same point b appears on
the left edge (Fig. 4.16), as well as on
the right edge of the planar model.
In fact, all points on an edge of this
rectangular planar model have their
corresponding points on the opposite
edge.

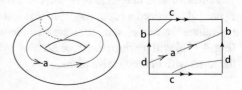

Fig. 4.16 Trajectory on a torus

A continuous path on the torus that starts and ends at point a (Fig. 4.16) is
represented in the planar model by a set of lines that go from point a to point b
from point b to point c and from c to d ending at a.

Thus, the planar model of the torus is a representation that maintains some
aspects of the surface geometry. We lose the usual geometric information, such as
lengths, areas, etc. However, relevant information to the study of the surface shape
is maintained.

The number of cuts needed to flatten a given surface depends on how complicated
the surface is.

Example 2 Sphere planar model

The spherical surface is so simple that it can be flattened with a single cut, for
example, along a segment of a meridian (Fig. 4.17).

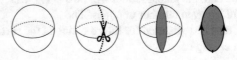

Fig. 4.17 Sphere planar model

To obtain planar models of more complicated surfaces will require more cuts. If
it takes n cuts on the surface to obtain a planar model, then it will be a polygonal
figure whose boundary is made of $2n$ edges; that is, twice the number of cuts.

Example 3 Bitorus planar model

It takes four cuts to flatten the bitorus (Fig. 4.10). Therefore, its planar model is an octagon.

To see this, we consider the planar model of the torus from which a disk (dotted line in Fig. 4.18) has been removed.

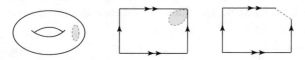

Fig. 4.18 Torus with a hole planar model

We flatten the torus using two cuts. We choose these cuts so that the boundary left by the removed disk passes through one of the corners of the rectangle. Then a cut through that vertex turns that boundary into a dotted edge. Consider two copies of the planar model of the torus from which a disk has been removed and identify them by the dotted edge (Fig 4.19).

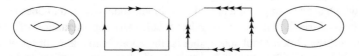

Fig. 4.19 Connected sum planar model

The first copy of the torus without a disk has a planar model whose edges have one and two arrowheads. The second copy has edges with three and four arrowheads. After identifying the dotted edge, we obtain the planar model of the bitorus, an octagon with pairs of edges with one, two, three and four arrowheads (Fig. 4.20).

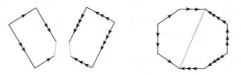

Fig. 4.20 Bitorus planar model

As each cut gives rise to a pair of edges, which must be indexed according to the direction of the cut, then the planar model of more complicated surfaces ends up with edges with an undesirably large number of arrowheads.

To simplify the planar models, we shall leave all pairs of edges with a single arrowhead, which will define the direction of the cut, and assign a letter to each pair of edges, which will indicate the pairs to be identified.

Therefore, planar models of any closed surface will be polygons with $2n$ edges, each pair of edges indexed by the same letter (pairs that will be identified in the reconstruction) and all edges will have a single arrowhead, which determines the direction of the cut.

When reconstructing the surface from the planar model, the direction of each edge is important. We have already seen that, in the planar model of a torus, if we invert the direction of one of the edges (in the Fig. 4.21, the direction of the right vertical edge is inverted), the identification of these edges generates a new surface.

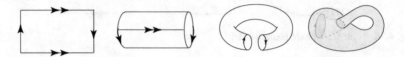

Fig. 4.21 Klein bottle planar model

After identifying the horizontal edges, two circles appear with opposite orientations, different from those obtained in the torus reconstruction. In order to identify these circles in a three-dimensional space with the indicated directions, the surface needs to be penetrated, identifying them from the inside. The result is the Klein bottle, which is an example of a closed two-dimensional object that does not embed into a three-dimensional space. In other words, any representation of the Klein bottle in a three-dimensional space; that is, the Klein bottle as a subset of a three-dimensional space, will have self-intersections. Only in spaces of dimension greater than three is it possible to embed the Klein bottle. However, its plane rectangular model is a very simple representation. This example illustrates the profound reach of planar models as a way of representing closed surfaces.

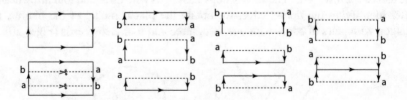

Fig. 4.22 Klein bottle as connected sum of two projective planes

Cutting and gluing operations on planar models enable us, in addition to simplifications, to obtain important information about the represented surface. As an example, we will transform the planar model of the Klein bottle into two Möbius strips models (Fig 4.22).

Initially, we cut the planar model of the Klein bottle along two parallel dotted lines. From this operation, we obtain a central Möbius strip and two other parts, which after being identified will form another Möbius strip. Finally, by identifying the two strips along the dotted edge of each, we obtain the Klein bottle.

These pictures (Fig. 4.22) constitute an alternative proof that the Klein bottle is the connected sum of two projective planes.

The cutting and gluing operation on planar models can also create alternative planar models of the same surface. For example, the two models in Fig. 4.23 represent a Klein bottle.

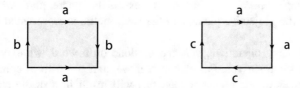

Fig. 4.23 Two planar models of a Klein bottle

In fact, cutting along the diagonal, you split the model into two parts, one upper and one lower, and create a new edge c (Fig 4.24). Then, the bottom is reflected horizontally to identify edges b of both parts. The result is another model of the same surface.

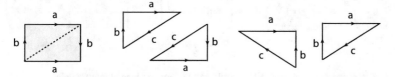

Fig. 4.24 Equivalent planar models

The procedure for obtaining planar models of two-dimensional objects can be adapted to three-dimensional objects. While in the two-dimensional case, the cuts generate edges (one-dimensional object), in the three-dimensional case the cuts generate faces (two-dimensional object) and the result will be a polyhedron with pairs of faces to be identified. In Chap. 7, we will represent some three-dimensional objects without boundaries using polyhedral models in three-dimensional space. Thus, we will be able to appreciate a little of the shape of those objects without resorting to spaces of dimension greater than three.

4.1.7   Word Representation of Surfaces

From a planar model of a surface, with its pairs of edges indexed by letters and the direction of cut determined by one arrowhead, a new representation of the surface can be created. This representation, called *representation by words,* is constructed by going along the boundary of the planar model starting from a vertex and following one of the two directions, clockwise or counterclockwise. When encountering a letter, if the direction of travelling is the same as the arrow, we simply collect the letter; otherwise, if the direction of travelling is opposite to the arrow, we collect the letter with exponent -1. At the end of the circuit, we will have collected all pairs of letters, with or without an exponent -1. This set of letters is called a word of the surface.

Some examples: aa^{-1} is a word that represents the sphere, $aba^{-1}b^{-1}$ represents the torus, aa the projective plane, and the Klein bottle is represented by the word $aba^{-1}b$.

There are many operations that can be done on a word, without it failing to represent the same surface. For example, if we start the path to collect letters at different vertices of the planar model, this will result in a cyclic change of the letters, hence a different word for the same surface. The same happens by reversing the direction of travel. For example, the torus $aba^{-1}b^{-1}$ is also represented by $b^{-1}a^{-1}ba$. Therefore, there are several synonyms for a word of a given surface (Fig. 4.25).

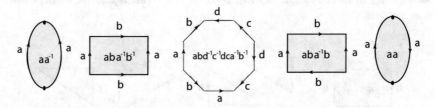

Fig. 4.25 Sphere, torus, bitorus, Klein bottle and projective plane planar models

Figure 4.23 above depicts two planar models of the Klein bottle. The word extracted from the one on the right is aacc. Recall that aa and cc both represent projective planes. This always occurs in the connected sum: a word representing the sum $F_1\#F_2$ is the concatenation of words of the surfaces F_1 and F_2. The bitorus, for example, has the word $b^{-1}a^{-1}bac^{-1}d^{-1}cd$, concatenation of the words of two tori.

In a word representation, there may still be room for simplification. In fact, words are taken from planar models, which in turn are created through cuts. If when flattening a surface we make redundant cuts, the corresponding word will contain redundant parts, which can be simplified.

A possible simplification is the following: if in a word of a given surface there is the sequence of letters aa^{-1}, which represents the sphere, it means that we are representing the connected sum of that surface and a sphere. Since the sphere is the neutral element of the connected sum, then the sequence aa^{-1} can be eliminated from the word. For example, the words $aa^{-1}bb^{-1}$ and aa^{-1} represent the same surface, because in the first case we have the connected sum of two spheres which, topologically, is the same as a sphere represented only by aa^{-1}.

A non-trivial and very useful operation for simplifying surface words is the following: if in a word a letter, say b^{-1}, lies between two equal letters, say a, then b^{-1} can be shifted by changing its exponent. In other words, $ab^{-1}a = baa$. We call this the *fundamental operation*. Moreover, $ab^{-1}b^{-1}a = bbaa$.

We will prove that $ab^{-1}a = bcc$.

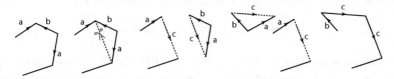

Fig. 4.26 Proof of the fundamental operation

Recall that cutting and gluing on planar models of surfaces are permissible operations in topology. We start with a planar model that has the sequence $ab^{-1}a$ in its word (clockwise). We cut the model along the dotted line (Fig. 4.26). This gives rise to a pair of edges that we will indicate with the letter c.

Then, we apply a reflection to one of the parts in order to identify the pair of edges indicated by the letter a.

The resulting sequence is bcc (clockwise). Hence, $ab^{-1}a = bcc$. As the letter a is no longer used after the operation, we can replace the letter c with the letter a. Thus, $ab^{-1}a = baa$ and the fundamental operation is proved.

Let us, for example, apply the fundamental operation to the Klein bottle word $aba^{-1}b$. As a^{-1} is between two letters b, we can shift it by inverting the exponent sign. Therefore, $aabb$ is a word for the Klein bottle.

As my colleague Azael Rangel Camargo, a professor at the *Instituto de Arquitetura e Urbanismo* (IAU-USP) would say: *the representation of surfaces by words is the utmost in synthesis.*

4.2 Topological Classification of Surfaces

On October 3, 2016, daily newspapers from several countries announced the awarding of the Nobel Prize in physics to a trio of British scientist experts in materials: David J. Thouless, F. Duncan M. Haldane and J. Michael Kosterlitz. The work, entitled *Topological phase transitions and topological phases of matter,* describes certain exotic states of matter (in addition to solid, liquid and gas) that occur at extreme temperatures, and their work uses the topological classification of surfaces that we will present below.

The topological classification of closed surfaces is ascribed by some authors to H. R. Brahana (1895–1972), in his Princeton PhD thesis in 1920 [1]. Others attribute it to M. Dehn (1878–1952) and P. Heegaard (1871–1948) [2].

The list of orientable surfaces starts with the sphere, followed by the torus and the connected sums of n copies of tori, called n-torus and denoted by nT^2

$$S^2, T^2, T^2\#T^2, \cdots, T^2\#T^2 \cdots \#T^2 = nT^2$$

The list of non-orientable surfaces starts with the projective plane, followed by the connected sums of n copies of projective planes, denoted by nP^2

$$P^2, P^2\#P^2, \cdots, P^2\#P^2 \cdots \#P^2 = nP^2$$

Using the fundamental identity $T^2\#P^2 = P^2\#P^2\#P^2$, we can simplify the connected sums of n projective planes. In fact, adding one P^2 on each side of the fundamental identity yields $T^2\#P^2\#P^2 = P^2\#P^2\#P^2\#P^2$. Hence, the sum of 4 copies of P^2 is equal to the sum of T^2 with the Klein bottle K^2; that is, $P^2\#P^2\#P^2\#P^2 = T^2\#K^2$. Now, adding one P^2 to each side of this last equality, we obtain $5P^2 = T^2\#K^2\#P^2 = T^2\#3P^2 = T^2\#T^2\#P^2$.

Successively repeating this process, we arrive at the following equalities:

$$(2n + 1)P^2 = nT^2\#P^2$$

and

$$(2n + 2)P^2 = nT^2\#K^2;$$

that is, the sum of an odd number $2n + 1$ of projective planes equals the sum of n tori with a projective plane and the sum of an even number $2n + 2$ of projective planes equals the sum of n tori with a Klein bottle.

4.3 Surface Identification

Having the list of all closed surfaces obtained from the topological classification, we can identify any closed surface. The procedure consists of obtaining a planar model and reading the word that represents the surface. In the case of closed non-orientable surfaces, whose models in three-dimensional space will always present self-intersections, we will remove one or more points in order to embed the representation of the surface in three-dimensional space; this will make it easier to choose cuts to obtain a planar model.

Let us look at some examples of this procedure.

4.3.1 Tripartite Unity

The sculpture made by the Swiss architect Max Bill (1908–1994), called Tripartite Unity, received the first prize at the first Bienal de Artes in São Paulo, in 1951 (Fig. 4.27). It is made of a thin metal plate, whose thickness is so small compared to its length and width that we can imagine the piece represents a surface.

Fig. 4.27
Max Bill 1951

Max Bill was fascinated by the Möbius strip and it seems that this sculpture contains one of those strips. It is easy to see that the boundary of the piece is just a closed line. So, we have a surface with a circle as the boundary. But, which surface does the sculpture represent?

After a few cuts, we will create a planar model of the surface and then we can read its word.

Four appropriate cuts will be needed to obtain a planar model. Indeed, we can divide the surface that represents the Tripartite Unity into two parts, with three cuts: one inferior and one superior (Fig. 4.28).

Fig. 4.28 Tripartite Unity planar model, step 1

The upper part of the surface is easily flattened (Fig. 4.29). It is continuously deformed into a flat object with three edges, each corresponding to one of the three cuts.

Fig. 4.29 Tripartite Unity planar model, step 2

The three edges of this flat object are indicated by the letters a, b and c (Fig. 4.30). These three edges are connected by arcs that represent part of the surface boundary.

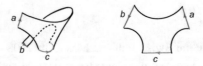

Fig. 4.30 Tripartite Unity planar model, step 3

The lower part of the Tripartite Unity requires one more suitable cut to be flattened (Fig. 4.31). It has three edges a, b and c of the first three cuts and two edges d corresponding to the last cut, in addition to arcs that represent part of the surface boundary.

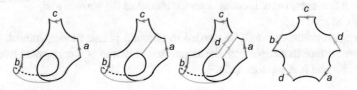

Fig. 4.31 Tripartite Unity planar model, step 4

Finally, we identify the two flat pieces, top and bottom, by one of the edges. We chose, without loss of generality, edge b (Fig. 4.32). Thus, we obtain a planar model of the surface that represents the Tripartite Unity.

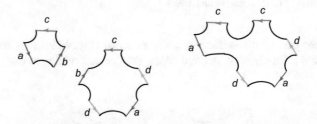

Fig. 4.32 Tripartite Unity planar model, final step

To identify the surface that represents the Tripartite Unity among all the surfaces topologically classified, we just have to read the word of the obtained planar model. Going along the boundary of the planar model starting at edge a on the left side and moving counterclockwise, we obtain the word $ada^{-1}dcc$. We can simplify this word using the fundamental operation. In fact, the letter a^{-1} is between two letters d and therefore can be shifted by inverting the exponent. As a result, the word for the Tripartite Unity is $aaddcc$.

Hence, the surface represented by the Tripartite Unity is topologically equivalent to the connected sum of three projective planes. It sounds like a good justification for the sculpture's name. Did Max Bill know that? It is easy to see that the sculpture represents a surface with a single closed line as the boundary. Therefore, the Tripartite Unity represents the surface obtained from the connected sum of three projective planes, from which a disk was removed.

4.3.2   An Olympic Surface

We give this name to the surface depicted in Fig. 4.33, as it resembles the logo of the Rio 2016 Olympic Games. This surface will be divided into three parts, using five cuts. Each part will be easily flattened.

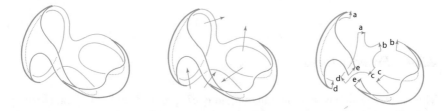

Fig. 4.33 Olympic surface planar model, step 1

To obtain a planar model, we initially identify the first two parts (Fig. 4.34) by edge d. The result is identified with the third part, by edge b. Finally, we obtain a planar model for the Olympic surface.

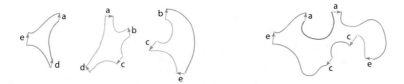

Fig. 4.34 Olympic surface planar model, final step

We read the word $ea^{-1}aecc$, then we cancel the sequence aa^{-1} and obtain $eecc$. As ee and cc represent projective planes, then the Olympic surface represents a Klein bottle with boundary. Going along the boundary of the surface, we find three closed lines. Therefore, the surface represents a Klein bottle with three circles as the boundary.

4.3.3 Surfaces with a Quadrilateral Planar Model

The sphere S^2, the projective plane P^2, the torus T^2 and the Klein bottle K^2 are the only closed surfaces whose planar models are quadrilateral figures (Fig. 4.35).

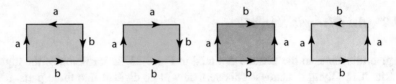

Fig. 4.35 Quadrilateral planar models

4.3.4 Francis Torus

This surface, which is important in our geometric proof of the fundamental identity $T^2\#P^2 = P^2\#P^2\#P^2$, represents a torus from which a disk has been removed. In other words, a torus with a single circular boundary.

We have already seen this surface when removing a disk from a torus (Fig. 4.11). We then deform the hole left by the removed disk and cut the surface. Finally, we deform and identify the edges created by the cut.

Here, we will do the reverse to identify the surface. We create a planar model with two cuts and we read the word of the Francis torus.

Fig. 4.36 Francis torus planar model

With the planar model, we obtain the word $aba^{-1}b^{-1}$. Based on this reading, we conclude that the surface represents a torus. The arcs of the planar model (Fig. 4.36) correspond to the boundary of the surface, and in the reconstruction they come together forming a closed line. Therefore, the Francis torus represents the surface of a torus from which a disk has been removed.

4.3.5 Surfaces with a Planar Model with Six Edges

Consider a planar model with six edges that define three identifications. Depending on whether the identifications are direct or else preceded by a 180° rotation, we will obtain distinct surfaces, as well as different numbers of boundary components. In the example (Fig. 4.37), assigning the letters a, b, c to each of the pairs of edges, the word obtained will be $abccba$.

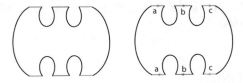

Fig. 4.37 Six edges planar model, type 1

Since words allow cyclical shifting of letters, this word is the same as $aabccb$. Applying the fundamental operation to the sequence, we obtain the equivalent word $aac^{-1}c^{-1}bb$. Therefore, the surface represented by this planar model is the connected sum of three projective planes $P^2\#P^2\#P^2$.

After going along the boundary of the planar model, we conclude that the boundary is a single circle. Therefore, this surface is topologically congruent to the Tripartite Unity of Example 1.

Keeping the upper arrows and taking all possible combinations of directions of the lower three edges, the surfaces represented will be of three topological types.

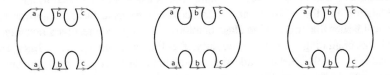

Fig. 4.38 Six edges planar model, type 2

Of the first type (Fig. 4.38), each surface has two boundary components and are represented by the words $abcc^{-1}ba$, $abccb^{-1}a$ and $abccba^{-1}$. The reader can check that the three surfaces are Klein bottles with two disks removed. Use $abccb^{-1}a = acb^{-1}cb^{-1}a = acc^{-1}b^{-1}b^{-1}a$.

Of the second type (Fig. 4.39), we have projective planes, each with three boundary components and represented by the words $abcc^{-1}b^{-1}a$, $abcc^{-1}ba^{-1}$ and $abccb^{-1}a^{-1}$.

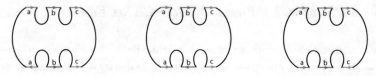

Fig. 4.39 Six edges planar model, type 3

Finally, the surface whose planar model has all six arrows pointing left to right has the word $abcc^{-1}b^{-1}a^{-1}$. Therefore, it represents a sphere and has four boundary components.

4.4 Shape of Objects

When we talk about the shape or format of an object, we are referring to its geometric appearance invariant by homeomorphisms. In other words, when two objects are topologically equivalent; that is, homeomorphic, we say that they have the same shape. In the case of two-dimensional objects, as already mentioned, the complete list of formats has existed since 1920. However, for three-dimensional objects we still do not have a classification of all formats, but it will not take long.

We call three-dimensional objects finite in size and without boundary as *hypersurfaces*.

Visualizing hypersurfaces and appreciating their shape is a much more difficult problem than the surface case. In fact, hypersurfaces live in high-dimensional spaces, at least one dimension more than our physical world. Thus, our perception of phenomena that occur in such esoteric spaces has to be expanded to have some understanding of the geometry and classification of hypersurfaces. Moreover, some technical issues that arise in the classification have only been resolved recently.

At the beginning of the last century, the French mathematician Jules Henri Poincaré (1854–1912) asked a question related to the geometry of three-dimensional objects without boundary. The question became known as the Poincaré Conjecture. The topological classification of hypersurfaces depends on the solution of the Poincaré conjecture. In addition to the conjecture, Poincaré left other notable problems whose solutions marked the development of contemporary mathematics.

One of those problems has an interesting story. In 1887, King Oscar II of Norway and Sweden created a competition, with a cash prize, to choose the best mathematical work on the stability of our solar system, a problem related to the so-called three-body problem, known since Newton's time.

Poincaré won the competition, but two years later he himself found an error in his winning work. The mistake was that he had assumed that certain complicated behavior did not exist. He had discovered what is now known today as *chaotic behavior*, something very complicated that arises as a result of successive simple events.

Fig. 4.40 Poincaré

The scandal of having the original work published in the mathematics journal *Acta Mathematica*, the most renowned at the time, was minimized by the publication of a new work to correct the previous one. The journal's editor, however, demanded that Poincaré not even mention the mistake made and also he should pay the costs of the new publication. This amounted to almost double the award received, which in the end helped the young Poincaré to become internationally known.

Poincaré became one of the most famous mathematicians of his day. According to French mathematician Étienne Ghys, Poincaré was very popular, so popular that his photo appeared in a collection of cards created by a chocolate manufacturer, which bore the image of the most famous people of the time. *How many mathematicians were so famous in life that their photo came out on candy bars?*—asks Ghys [4, p. 89] (Fig. 4.40).

Generations of brilliant mathematicians searched in vain for a solution to the Poincaré conjecture, until in 2006 it was finally solved by the eccentric Russian mathematician Grigori Perelman (Fig. 4.41). For this spectacular achievement, which should help describe the shape of our universe, Perelman was chosen to receive the Fields Medal, one of the most prestigious awards in mathematics. However, Perelman declined the distinction and also declined the 1 million US dollar prize that an Institute offered to anyone who could solve the Poincaré conjecture.

Fig. 4.41 Perelman

The topological classifications of objects according to their dimension ends here; that is, the problem in dimensions one and two are solved, dimension three is on its way, but dimension greater than three is unsolvable. This is the work of the Russian mathematical logician Andrei A. Markov Jr. [5].

References

1. Brahana, Henry; *Systems of circuits on two-dimensional manifolds,* Ann. of Math. (2) 23, 144–168 (1921).
2. Dehn, M.& Heegaard, P.; *Analysis situs,* in Enzyklop. d. math. Wissensch. III1, 153–220 (1907).
3. Francis, George; *A topological picturebook*, Springer 1990.
4. Ghys, Étienne; *A singular mathematical promenade*, ENS Éditions 2017. http://perso.ens-lyon.fr/ghys/promenade/
5. Markov Jr., Andrei; *The insolubility of the problem of homeomorphy,* Proc. International Congress of Mathematicians in Edinburgh, pp. 300–306, Cambridge Univ. Press 1958.
6. Rodríguez, José Luis; https://topologia.wordpress.com/2013/01/27/superficies-topologicas-en-el-arte/

Chapter 5
The Fourth Dimension

5.1 Flatland and Spaceland

The physical universe is the place of all known matter.

The three-dimensionality of this universe is knowledge obtained through the senses.

Any displacements around the world are combinations of displacements in three independent directions. The Cartesian description (Fig. 5.1), which associates three numbers to each point P in space; namely, length x, width y and height z, measured from a fixed coordinate system, creates a mathematical representation of the three-dimensional universe. This coordinate description $P = (x, y, z)$ is denoted by $\mathbb{R}^3$ and it is known as the Euclidean three-dimensional space. Analogously we define $\mathbb{R}^n$, for all n.

Fig. 5.1 Coordinates

The possible existence of spaces with dimension greater than three, of a fourth dimension for example, is the object of a very old debate.

Aristotle (385 BC–322 BC) denied any possibility of the existence of a fourth dimension, and after him many mathematicians and philosophers claimed to have a proof of this. According to the historian of mathematics Florian Cajori (1859–1930), the list of non-believers is long: C. Ptolemy (90–168), G. Leibniz (1646–1716), I. Kant (1724–1804), to name just a few famous people who indulge in loose ratiocinations, mixing God with the inverse-square law [3, p. 402].

In the Bible, in Ephesians 3:18, an extra dimension is mentioned: ...*may have strength to comprehend with all the saints what is the breadth and length and height and depth* ....

In physics it is common to decompose 4-dimensional spaces into $3 + 1$, three dimensions to locate a body and the fourth dimension to be interpreted.

© The Author(s), under exclusive license to Springer Nature Switzerland AG 2022
T. Marar, *A Ludic Journey into Geometric Topology*,
https://doi.org/10.1007/978-3-031-07442-4_5

One of the earliest records of this interpretation was made by Henry More (1614–1687), a Cambridge University theologian, a contemporary of Isaac Newton. According to More, the fourth dimension is the place of the soul [3, p. 399].

The most popular interpretation of the fourth dimension was introduced by the Frenchmen D'Alembert (1717–1783) and Lagrange (1736–1813) who, in their mathematical descriptions of problems in mechanics, associate the fourth dimension with time.

In geometry there is nothing spectacular about the idea of spaces of dimensions greater than three. These spaces are as abstract as one-, two- or three-dimensional spaces. What may be surprising is that geometric representations in high-dimensional spaces, which transcend our ability to visualize, phenomena that are alien to our experience of the physical world may occur.

The classic book *Flatland* by Edwin A. Abbott (1838–1926) [1] describes the life of beings that inhabit a two-dimensional universe, called *flatlanders*. The difficulty that flatlanders experience to understand the third dimension, which they consider esoteric, is analogous to our difficulty with the fourth dimension. This analogy help us to understand certain phenomena that take place in spaces of dimension greater than three.

For example, in a two-dimensional world, as in Flatland, edges define a closed polygon and this encloses a two-dimensional place, its interior. Flatlanders can enter that place, moving one of the edges like opening a door. Everything takes place on the plane (Fig. 5.2).

Fig. 5.2 Flatland

We, who live in three-dimensional space, can jump into that two-dimensional place without opening the door. The closed place of the plane is open in the third dimension (Fig. 5.3a).

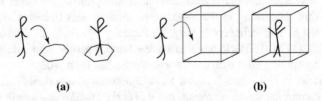

(a)   (b)

Fig. 5.3 Jumping inside

Let a solid square be a plane quadrilateral together with its interior points. Then, six solid squares can be arranged to define a cube, with a floor, ceiling, and four vertical walls. The interior of the cube encloses a three-dimensional place, a portion of the three-dimensional space. We can enter that place through an open door in one of the cube walls.

By analogy with what happened to the closed two-dimensional place, which is open in the three-dimensional space, the closed three-dimensional place is open in the four-dimensional space. By making use of the fourth dimension, it is possible to enter that three-dimensional place without needing to open the door. Figure 5.3b depicts a jump through the fourth dimension into a closed three-dimensional place defined by transparent solid squares.

There is no magic portal from one world to another of higher dimension. For example, a flatlander cannot enter the three-dimensional place defined by a cube that rests on Flatland. In fact, that inhabitant sees nothing of the cube beyond the ground, which is the only part of the cube contained in the two-dimensional world. Even if the access door to the three-dimensional place is open, the inhabitant of the two-dimensional world does not find a place to move into, because the floor is a solid square.

In the romance by Edwin Abbott, an inhabitant of the third dimension, a sphere, takes a flatlander, a square, for a trip into the third dimension. That was a transcendental experience for the flatlander.

Just as one enters a three-dimensional place without opening the door, using the fourth dimension, we can likewise leave a closed room, as it is open from the point of view of the fourth dimension. Everything that is closed in the three-dimensional space is open in the four-dimensional one. By using the fourth dimension, an egg yolk can be removed without breaking the egg; that is, you *can* make an omelette without breaking eggs using the fourth dimension. From the fourth dimension, the interior of any closed region of a three-dimensional space can be seen, such as our body. A surgeon from a four-dimensional space can perform invasive surgery without cutting the patient. Money from a locked vault can disappear without being broken into!

5.2 Beyond the Third Dimension

We are going to explore geometric representations beyond the third dimension. This is the title of a very nice book by a leader in the study of higher dimensions, Thomas Banchoff [2], which investigates ways of picturing and understanding dimensions below and above our own.

Aside from esoteric issues, high-dimensional spaces can be used to simplify representations of phenomena that are normally complicated when considered in lower-dimensional spaces.

For example, in the topological classification of two-dimensional objects, the very ones that represent the possible shape of Flatland, we have seen that the models of non-orientable closed surfaces in a three-dimensional space can be very complicated. In those cases, a space of dimension greater than three is necessary for a full understanding of the model making process. Indeed, all orientable surfaces, such as the sphere (Fig. 5.4a) and the torus (Fig. 5.4b), embed in a three-dimensional space; while, the non-orientable ones, such as the projective plane (Fig. 5.4c) and the Klein bottle (Fig. 5.4d), need an extra dimension to be represented without self-intersections.

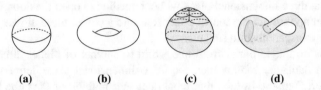

(a)   (b)   (c)   (d)

Fig. 5.4 Sphere, torus, projective plane and Klein bottle

All closed surfaces, orientable or non-orientable, embed in a four-dimensional space and any projection of these objects into a three-dimensional space can present self-intersections, and in some cases singular points, as well. What assures us that all surfaces embed in a four-dimensional space is a theorem by Hassler Whitney (1907–1989). He proved that n-dimensional objects embed into $2n$-dimensional environments, for any $n \geq 2$. Thus, any surface ($n = 2$) embeds into the four-dimensional space.

The condition of a space not being ample enough to contain a representation without irregularities also occurs in the representation of projections of one-dimensional objects in two-dimensional spaces. For example, a curved and twisted line (one-dimensional object) will certainly embed in a three-dimensional space, but when it is projected onto the plane (two-dimensional space) it may have irregularities such as self-intersections (crossings) and singular points (cusps), depending on the projection direction. The cusp occurs exactly when the projection direction is tangent to the curve. The point of tangency of the curve is projected to the plane as a cuspidal point (Fig. 5.5).

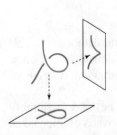

This phenomenon was physically introduced to me by my then Ph.D. supervisor at Warwick University (UK), David Mond. Moving a twisted wire, as I watched it with one eye closed, David made me see that at a certain moment, the succession of crossings projected on the surface of my retina, a two-dimensional space, gave way to a cuspidal point.

Fig. 5.5 Plane projection

Let us consider spaces of a dimension greater than three and embed any surface into them. Once embedded in the four-dimensional space, we can represent, and perhaps see the closed surface in its fullness. This process of seeing a closed surface embedded into the fourth dimension, and not just their complicated projections onto a three-dimensional space, bears a certain resemblance to what occurs in Plato's allegory of the cave.

All procedures that make use of a four-
dimensional space are quite abstract, as we are not
used to seeing spaces beyond our three-dimensional
physical world. Also some representations carry certain dubiety and the observer must choose what it is
seen.

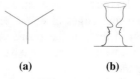

(a)   (b)

Fig. 5.6   Optical illusion

In Fig. 5.6a, do we see a corner of a room or a
vertex of a cube?; that is, the intersection of the three
segments is backward or forward? In Fig. 5.6b, do we see two profiles or a vase?
The observer chooses what is seen and only sees one of the choices.

The magnificent Brazilian artist, Regina Silveira [7], is the author of works that
play with our senses. We are taken by our visual perception beyond the space of
representation. Figure 5.7 shows Regina Silveira standing on her work made on the
floor, but it looks as if she is floating.

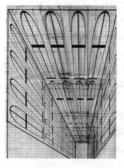

Fig. 5.7   Regina Silveira

Similar to the situation in which we carry a flatlander into a three-dimensional
place, we can imagine ourselves in the same situation, in a higher dimension space.
If we conceive a place in a four-dimensional space and are carried into it, perhaps by
some inhabitant of that space, we will certainly have a transcendental experience.

5.3   Four-Dimensional Place

In Chap. 3, we mention the notion of place, space and matter described by Archytas
in ancient Greece. Archytas asserts that space is where all phenomena occur, but
the notion of *place* is primordial. According to him, a body occupies a place and
cannot exist without it. Furthermore, moving a body requires having a place for it
to occupy.

To represent a place in four-dimensional space we use analogy with the process
of modelling places in lower-dimensional spaces.

It is easy to represent a square in the plane of this page, with its four right angles and four edges of equal length. A square, as all polygons, is a one-dimensional object that, in the plane, encloses a region of a two-dimensional space, a two-dimensional place. We have called the square together with its interior points *a solid square*.

The representation of a cube on this same page is not so straightforward, although very familiar, it contains several agreements between what is represented and what is to be seen. In fact, a set of three segments two by two orthogonal, as it occurs at each vertex of a cube, is only possible to represent in the plane with the help of certain artifices. Using perspective, we represent on the plane of this page the image that is projected on the surface of our retina when we look at the surface of a cube. Perspective distorts the figure. Some or all of the cube's square faces are transformed into other types of quadrilaterals when plotted on the plane. Such distortions are intellectually compensated for through a certain agreement between what one sees and what it actually is. Therefore, we manage to see in the two-dimensional plane an object of the third dimension.

A cube as a closed two-dimensional object is the set of 8 vertices, 12 edges and 6 solid square faces. These faces are 3 pairs of parallel solid squares, two for each of the three directions defined by the cube. Thus, this closed two-dimensional object encloses a region of the three-dimensional space, a three-dimensional place. Analogous to the solid square, we call a cube together with its interior points *a solid cube*.

To draw a cube on the plane, we draw three segments in perspective, say s_1, s_2 and s_3, with a common extremity (Fig. 5.8). This will represent a vertex and three edges of the cube, therefore, in reality, they are two by two perpendicular. These three segments have three end points. The end of one of the segments must receive two segments parallel to the other two. For example, the end of segment s_1 receives two segments, one parallel to s_2 and the other parallel to s_3. This also creates new end points and we repeat the process until all eight vertices have three edges (Fig. 5.8).

Fig. 5.8 Cube

Note that, since the beginning of the drawing, we made agreements between what is drawn and what is represented. The three leading edges represent two by two perpendicular edges. Since, at most, two perpendicular edges pass through a given point in the plane, then the three edges are drawn using perspective. The choice of edge directions determines the point of view from which the cube will be perceived when represented in the plane.

Similarly, in order to define a four dimensional place, we start with four segments with common extremities at a vertex, each of which represents a direction perpendicular to the other three (Fig. 5.9). Then, we do the same at the four ends created, drawing three segments, each one parallel to one of the initial edges. We repeat the process until we have four of those edges at each of the 16 created vertices.

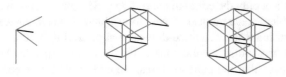

Fig. 5.9 Hypercube

As a result, we obtain a representation of the 16 vertices and 32 edges of the object called hypercube (or tesseract, from the Greek four rays). Analogously to what we did to define the closed two-dimensional object cube, adding the 6 solid squares, here we have to consider the 8 solid cubes to define the closed three-dimensional object hypercube. Therefore, this closed three-dimensional object encloses a region of the four-dimensional space; that is, a four-dimensional place.

Although the representation of the hypercube in the plane follows a simple procedure, as in the drawing of the cube, visualization is much harder. While the geometric object cube is familiar to us, since it can be represented by physical models, the hypercube has no physical model in a three-dimensional space.

There is another technique to produce representations of squares, cubes and hypercubes. This is done dynamically by sliding.

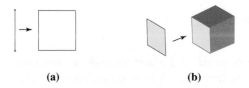

(a)   (b)

Fig. 5.10 (a) From 1D to 2D   (b) From 2D to 3D

The square is obtained by sliding an edge along a direction perpendicular to that edge. The edge of length L is slid, following a perpendicular direction, by the same length L. In the end we will have a new edge of length L and, between the start and end edges, we collect only the two extreme points of the sliding edge (Fig. 5.10a). The result is a square, one-dimensional object, which encloses a two-dimensional region of the plane defined by the direction of the starting edge and its perpendicular.

The cube can also be obtained as the result of sliding a solid square in the direction perpendicular to the containing plane, forward or backward, by the same length as the edge of the square. While in the plane, the two perpendicular directions allow movement from left to right and from top to bottom, a back and forth movement is only possible in a three-dimensional space. We start with a solid square, with its four vertices, four edges of length L and the two-dimensional region defined by them, which we call a *face*. We slide this face by length L along the direction perpendicular to the plane containing the starting face. At the end of this action, we will have a new face, parallel to the initial one and during the sliding process, we collect only the edges of the sliding solid square. Thus, we obtain a two-dimensional object made up of six faces (solid squares), which encloses a region of the three-dimensional space, a three-dimensional place (Fig. 5.10b).

Similarly, when sliding a solid cube in the direction perpendicular to the three-dimensional space that the cube encloses, we end up with a hypercube, a closed three-dimensional object. It is an identical process to the two previous ones that generated a square (closed one-dimensional object) and a cube (closed two-dimensional object), however the perpendicular direction to the space defined by the cube cannot be materialized in a three-dimensional space. We start with a solid cube with edges of length L; that is, the cube and all the points of the three-dimensional region that it encloses. We slide this solid cube in the direction perpendicular to the three-dimensional space that contains it and, at the end of the sliding of length L, we have another solid cube (Fig. 5.11a). In the sliding process, we collect only the faces (solid squares) of the solid cube. As a result of this action, we get a hypercube (Fig. 5.11b)

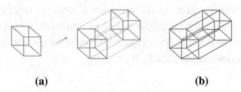

(a)   (b)

Fig. 5.11 From 3D to 4D

Corresponding to the six solid squares (two-dimensional object) that make up the faces of the cube, the hypercube is formed by eight solid cubes (three-dimensional object), two for each of the four directions that define it and they are called *cells* (Fig. 5.12)).

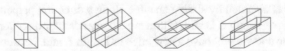

Fig. 5.12 Hypercube cells

Following the Lagrangian interpretation of four-dimensional space; namely, physical space together with time, our entire life could be represented by the sequence of three-dimensional places we occupy, one for each instant of our life, generating something similar to a hypercube.

In his 1912 *Nude descending a staircase*, Marcel Duchamp (1887–1968) suggests the representation of an event that takes place in an interval of time (Fig. 5.13).

If we add an extra dimension to the Lagrangian spacetime, we can represent another time, regardless of the first one. Therefore, in this five-dimensional space, different events of our life can be represented, simultaneously.

At the end of the movie *Interstellar* (2014), when the character Cooper returns to his home, this phenomenon occurs. He can appreciate different moments of his life, but he is not allowed to pass from one time to another. Cooper is confined to a hypercube, which represents only moments in his life. His daughter, as a child and as an adult, both appear simultaneously

Fig. 5.13 Duchamp 1912

in different situations, in two distinct hypercubes. A clear reference to the fifth dimension.

In the next chapter, the fifth dimension is required to understand a specific model of the projective plane represented in a three-dimensional space.

These n-dimensional spaces expand our geometric perception, at least theoretically. There are some artifices for understanding models that enclose regions of n-dimensional space using $(n-1)$-dimensional representations. Let us see this in the case of the cube and the hypercube represented, respectively, in the plane and in a three-dimensional space.

The flat representation of a cube is well known.

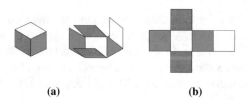

(a)   (b)

Fig. 5.14 Unfolding the cube

The flat model of the cube is obtained by cutting seven of its 12 edges (Fig. 5.14a. The result is the flattening of all six square faces, connected by the remaining five edges (Fig. 5.14b).

The composition of the six square faces with fourteen outer edges, created by the seven cuts and five edges connecting the faces of a cube, will be called a *cube unfolding*. Conversely, from a cube unfolding, pasting (identifying) suitable pairs of edges, we reconstruct the cube.

These plane representations of the cube have the advantage of not requiring perspective. Whereas in perspective drawing, square faces of the cube appear as rhomboids, in the unfolding none of the square faces are distorted.

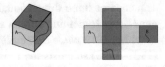

Fig. 5.15 Trajectory on the cube

Getting used to the identifications of the unfolding, it can create a geometrical idea of the surface. For example, a continuous path over the cube surface, from point *A* on one face to point *B* on another face and going through a third face, is represented by three disjoint curves in the unfolding (Fig. 5.15).

Several compositions with six square faces generate other cube unfoldings; you simply have to choose another set of seven edges to receive the cuts that will flatten the cube. However, not all six-square arrangement with five inner edges is a cube unfolding. There are exactly 11 cube unfoldings (Fig. 5.16).

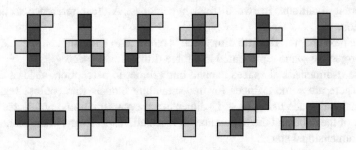

Fig. 5.16 The 11 cube unfoldings

This procedure of cube unfolding was adapted in Chap. 4 to construct planar models of any closed surface. Appropriate cuts were applied to models made of an imaginary material that was perfectly deformable, until the surface was flattened.

While the cube has 8 vertices (0-D), 12 edges (1-D) and 6 faces (2-D), the hypercube has 16 vertices (0-D), 32 edges (1-D), 24 faces (2-D) and 8 cells (3-D).

A three-dimensional model of the hypercube, analogous to the cube unfolding, is a composition of 8 three-dimensional cells (solid cubes) obtained by cutting along 17 square faces, out of the 24. Having these cuts, the hypercube unfolds into a three-dimensional model, a composition of 8 solid cubes. Such an unfolding is also called an *open hypercube* (Fig. 5.17).

Fig. 5.17 Open hypercube

Each cut along the squares results in a doubling of that square in the three-dimensional model. Of the 24 square faces of the hypercube, 7 faces do not receive cuts when making the three-dimensional model; they are the inner faces.

The 11 different cube unfoldings were obtained from different choices of the set of seven cut edges. The three-dimensional unfolding of the hypercube is also countable. There are exactly 261 different hypercube unfoldings in a three-dimensional space, for different choices of the 17 square faces to be cut. A proof, obtained in a combinatorial way using graphs, can be found in the article by P. Turney [8].

When rebuilding the cube from its unfolding, only one square face will remain on the plane where the unfolding rested. After properly identifying the pairs of edge, the other five square faces of the cube unfolding will occupy the third dimension.

The collages to reconstruct the hypercube from its three-dimensional model are made in the fourth dimension, identifying pairs of squares obtained from the 17 cuts. Identifying the squares, point by point, is done as if there were hinges on the edges of the cubes, and the base square is identified with the roof square. After identifying them, similarly to what happened to the cube, only one of the eight solid cubes that make up the open hypercube will remain in the three-dimensional space were it stood, the other seven solid cubes will occupy the fourth dimension.

5.4   The Fourth Dimension in Arts and Literature

We can find references to the hypercube in the work of some writers and artists.

In Robert Heinlein (1907–1988) short story —*And he built a crooked house*— of 1941, there is a futuristic three-story building beyond the ground floor. On the first floor, four bedrooms converge in a central room and two more rooms on the upper floors, all in the shape of an open hypercube.

The story takes place in California, where earthquakes are frequent. One day there was an anomalous earthquake. Rather than destroying the house, the earthquake described by Heinlein properly identified squares of the open hypercube, and thus the open hypercube-shaped house, closed!

After identifying them, seven cubic cells went to the fourth dimension, leaving only a single cube visible in the three-dimensional space, the very one in which the sliding entrance door was installed. The intact door invited the characters to appreciate the interior of the house. In that room on the ground floor, there was a staircase that led to the upper floor. They went up to the first floor and found a hall with doors leading to the four bedrooms. Inside one of these rooms, they opened the window and, passing through it, reached the top floor. On the ceiling of that same room, there was a passage to the second floor, and also a walkway on the floor that led back to the entrance hall on the ground floor. A very crooked house indeed.

Heinlein's tale is very interesting, but it lacks rigor. In fact, the eight cubes that could give rise to a hypercube, following an adequate identification of pairs of square faces, and hence defining a four-dimensional place, they are solid cubes and, therefore, impossible to be penetrated by humans from a three-dimensional space. When sliding the entrance door, we would be confronted with the solid contents of the cube. As Heinlein describes the house, its 8 cubes are not solid cubes. Thus, after properly identifying the square faces, they will not define any four-dimensional region. It is the same as imagining a flat model of the cube with only the edges, without the inner part of the square faces. Once the edges are properly identified, a closed region in a three-dimensional space would not be defined, only a set of 12 edges, of which three of them meet in each of the eight vertices. The flat cube model is made up of six two-dimensional square regions (solid squares), while the open hypercube is made up of eight three-dimensional cubic regions (solid cubes).

Just as an inhabitant of Flatland would need our help to enter a cube, we too, to appreciate the four-dimensional place limited by a hypercube, would have to be transported by a being from the four-dimensional world. There is no magic portal for us to get into a four-dimensional space.

Ignoring the technicalities, Heinlein's tale is excellent and all the square face identifications described in the tale to generate the hypercube are correct.

René Magritte (1898–1967), in his *La reproduction interdite* of 1937 [6], portrays a character who looks straight ahead in the mirror and sees his back, a phenomenon that Heinlein's characters could appreciate looking up inside the hypercube house. In fact, among the 17 identifications of the crooked house that generate the hypercube, the square face of the roof of the house is identified to the square face which is the entrance hall floor .

Salvador Dali (1904–1989) in 1954 crucifies Christ in an open hypercube, in his *Corpus Hypercubus* [4].

Some time ago, in religious education classes at Catholic schools, the following warning could be heard: it's no use hiding under the bed or locking yourself in the bathroom, *He sees you wherever you are*; in other words, any closed three-dimensional place is open to *Him*.

When crucifying the Christ in a higher dimension, perhaps Dali confesses that he had one of those religious education lessons in his childhood.

The fourth dimension will always be the subject of esoteric fantasies.

References

1. Abbott, Edwin; *Flatland*, introd. by Thomas Banchoff, Princeton University Press 2015.
2. Banchoff, Thomas; *Beyond the third dimension*, W H Freeman & Co 1990.
3. Cajori, Florian; *Origins of Fourth Dimension Concepts* , The American Mathematical Monthly 33, 397–406 (1926).
4. Dali, Salvador; https://www.metmuseum.org/art/collection/search/488880
5. Heinlein, Robert; *And he built a crooked house* , Street & Smith Publications 1941.
6. Magritte, René; https://www.moma.org/audio/playlist/180/2381
7. Silveira, Regina; https://reginasilveira.com/
8. Turney, Peter; *Unfolding the Tesseract,* Journal of Recreational Mathematics 17, 1–16 (1984).

Chapter 6
Non-orientable Surfaces

6.1 Model Making in Three-Dimensional Space

The topological classification of closed surfaces produced two lists (Chap. 4). The first list of surfaces starts with the sphere, followed by the torus and all connected sums of tori (Fig. 6.1), such as the bitorus, the tritorus, and so on. These are orientable surfaces.

Fig. 6.1 Sphere, torus, bitorus and tritorus

All closed orientable surfaces embed in three-dimensional space; in other words, we can model them in three-dimensional space without self-intersections. Furthermore, these surfaces divide the three-dimensional space into two regions: interior and exterior. Closed orientable surfaces are sometimes referred to as *two-sided surfaces*. To go from the interior to the exterior defined by a closed orientable surface, a hole needs to be drilled in the model. However, in four-dimensional space, the term two-sided is meaningless, as it is possible to penetrate the closed surface using the fourth dimension.

The other list of closed surfaces obtained in the topological classification is that of non-orientable surfaces. These are interesting two-dimensional objects because they dispute with our intuition.

Imagine the surface of a long road in the shape of a Möbius strip (Fig. 6.2). For someone moving on it, it feels like an ordinary path. However, there are trajectories that can make the hiker return to the starting point upside down. This trajectory

Fig. 6.2 Möbius strip walk

T. Marar, *A Ludic Journey into Geometric Topology*,
https://doi.org/10.1007/978-3-031-07442-4_6

is called a *disorienting path.* For this reason, we say that a Möbius strip is a one-sided surface. Furthermore, any surface that contains a disorienting path; that is, containing a Möbius strip is a non-orientable surface. In other words, the presence of a Möbius strip on a surface is equivalent to the non-orientability.

The list of closed non-orientable surfaces starts with the projective plane, which, as we have seen (Chap. 4), is represented by a Möbius strip with a disk glued along its boundary. Then, the list is completed with all connected sums of projective planes, such as the Klein bottle, which is the connected sum of two projective planes.

Models of closed non-orientable surfaces in three-dimensional space can be very tricky, as none of these surfaces embeds in three-dimensional space. The three-dimensional space is not ample enough to avoid the self-intersections.

Fig. 6.3 Projective plane and Klein bottle

For example, in three-dimensional space, the projective plane has a model with a self-intersection along a segment, while a Klein bottle model has a circle of self-intersections (Fig. 6.3).

In general, self-intersections occur along lines, a phenomenon that results from transversal intersection of two planes. It can also happen that, in three-dimensional space, three planes meet transversally and have a single point in common, called a *triple point,* similar to a corner of a cubic room.

In spaces one dimension above; that is, in the fourth dimension, all closed non-orientable surfaces can be embedded as the self-intersection can be undone.

This process of undoing self-intersections when adding an extra dimension to the ambient space is quite intuitive. For example, consider in a plane (a two-dimensional space), a pair of lines intersecting at a point. Small changes in the position of these lines in the plane maintain the intersection. However, if we add a dimension to the plane; that is, if we consider the concurrent lines

Fig. 6.4 Lifting

in three-dimensional space, we can lift one of them and undo the crossing (Fig. 6.4). Similarly, two planes that intersect along a line in three-dimensional space can be moved in the fourth dimension in order to undo the self-intersecting line. Do not try to see it, just imagine it.

Thus, in high-dimensional spaces, geometric representations can be simplified. For instance, all closed non-orientable surfaces embed in four-dimensional space, and therefore can be represented without self-intersections. While a model of the projective plane in three-dimensional space has a self-intersecting segment, in four-dimensional space each point of the self-intersection can be undone. It is a difficult procedure to visualize, as any representation in the fourth dimension.

Fig. 6.5 Undoing intersection

Some writers call the Klein bottle a one-sided surface, and claim it has no inside. According to Sir Christopher Zeeman, we should avoid the term one-sided for the Klein bottle or any other closed non-orientable surface. The Möbius strip is one-sided (it is not closed as it has a boundary). Zeeman argues using a lower dimensional example. Consider a figure eight curve in the plane (Fig. 6.5). One could talk about interior or exterior points of the curve in the plane, but if we embed that curve in three-dimensional space and undo the crossing, the inside and outside points no longer make sense. The same happens to the Klein bottle. Because of the unavoidable self-intersection in three-dimensional space, *an ant cannot crawl from one side to the other,* says Zeeman. Furthermore, the self-intersection can be undone in the fourth dimension, and again the inside and outside are meaningless. Therefore, Zeeman concludes: *the Klein bottle cannot truthfully be said to be one-sided in either three or four-dimensions. Consequently we prefer the term non-orientable to the term one-sided.*

From the topological classification we know that all closed non-orientable surfaces are equivalent to connected sums of a certain number of projective planes. We also know that the connected sum of two surfaces is obtained by removing a disk from each surface and identifying the boundaries created by removing the disks. Therefore, to create models in three-dimensional space of any non-orientable closed surface, it is sufficient to obtain models of the projective plane in three-dimensional space and do a connected sum with the appropriate number of copies of those models.

Recall that at the end of Chap. 3, we introduced the projective plane as the Euclidean plane together with all its points at infinity. In addition, we represented the Euclidean plane by an open disk; that is, a disk without its circular boundary, and the points at infinity as identified pairs of antipodal points on the circular boundary of the disk.

In Chap. 4, we mentioned the projective plane model obtained by pasting a disk along the boundary of a Möbius strip. This a process not so simple to visualize. However, we can work backwards, decomposing a projective plane in these two parts: a Möbius strip and a disk.

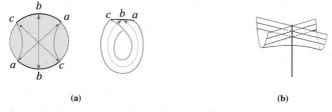

(a)   (b)

Fig. 6.6 (a) Projective plane   (b) Whitney umbrella

On identifying the antipodals of the circular boundary of the disk, a Möbius strip appears plus something else. In fact, the disk is divided into three parts: a central part corresponding to the Möbius strip and two lateral parts (Fig. 6.6a). Identifying these two lateral parts along the dotted arc between points *a* and *c* creates a disk. Therefore, a model of the projective plane is obtained from a Möbius strip and a disk, identified by their boundaries, point by point. This boundaries identification in three-dimensional space creates self-intersections. Indeed, if we travel along the boundary of the Möbius strip, we will go around twice until we return to the starting point, while the disk has a circular boundary that closes after one turn. To solve this, we will embed the Möbius strip into the fourth dimension and project it down to three-dimensional space in such a way that the boundary of the strip is transformed into a curve that closes after one turn. In this process, a self-intersecting segment appears.

In addition to self-intersections, models of closed non-orientable surfaces in three-dimensional space may present another irregularity, known as a *singular point*. This singularity of surfaces was found by Hassler Whitney in 1944 [11]. Due to its shape (Fig. 6.6b), it is called *the Whitney umbrella*.

The models of the projective plane, and therefore of all closed non-orientable surfaces, will be constructed assembling three types of basic modules; namely, the transversal intersection of two planes, the transversal intersection of three planes (triple point) and Whitney umbrellas (Fig. 6.7). All these pieces are made of a perfectly deformable material, as it is usual in the making of topological models. At the end, with pieces of plane we complete the model of the closed surface, as in a construction toy.

The transversal intersection of two planes occurs along a line (Fig. 6.7a). The intersection of three planes, two by two transversal, produces three lines that meet at a single point, called a triple point (Fig. 6.7b). The umbrella features a self-intersecting line with a singular point at one end (Fig. 6.7c).

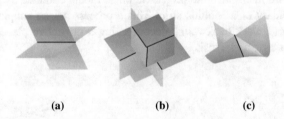

| (a) | (b) | (c) |

Fig. 6.7 Surface basic building blocks

Thus, when making models by combining these three basic modules, the self-intersections are also combined into a curve, called *double point curve*. Double point curves can have a finite number of triple points and, if it has endpoints, they will be singular points. The models of the projective plane and the Klein bottle (Fig. 6.3) have, respectively, a segment with two singular points and a circle as double point curves.

Note that the basic modules are made up of the double point curve and pieces of plane that rest on it. In the first module (Fig. 6.7a) there are four pieces of plane, in the second (Fig. 6.7b) 12 pieces of planes and in the umbrella (Fig. 6.7c) there are two curved pieces of plane attached to the double point segment.

Double point curves of models are similar to skeletons on which pieces of plane rest. A double point curve with a triple point and two singular endpoints is shown in Fig. 6.8a. Two pieces of plane are attached to it, as can be seen in Fig. 6.8b. The model of the surface is completed with pieces of plane (Fig. 6.8c) [6].

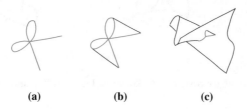

(a)   (b)   (c)

Fig. 6.8 Surface skeleton

Although the double point curve contains a great deal of geometric information about the surface, it does not quite determine it. In other words, there are models that have the same double point curve, but they represent topologically distinct surfaces in three-dimensional space.

For example, consider two Whitney umbrellas and identify the X-shaped parts of their boundaries. The result is a model whose double point curve is a straight line segment with two singular points at the extremities.

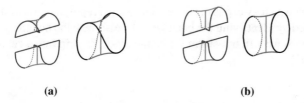

(a)   (b)

Fig. 6.9 Two ways to identify two Whitney umbrellas

There are two choices for identifying the X-shaped parts and the two resulting models are topologically distinct, yet they have the same double point curve. In fact, one of the models has a single curve as the boundary (Fig. 6.9a), while the other has two (Fig. 6.9b). In the first case, with one piece of plane we close the model, while in the second case two pieces are needed. Each of these closed models represents a surface in three-dimensional space, and they are not topologically equivalent. We will see that the first model represents a projective plane, while the second represents a sphere in which the points of the equator were identified in pairs, creating the segment of double points.

It can also happen that the same surface is represented by models whose double point curves are distinct.

We will describe in detail three classical representations of the projective plane in three-dimensional space, called *sphere with a cross-cap* (Fig. 6.10a), *Steiner Roman surface* (Fig. 6.10b) and *Boy surface* (Fig. 6.10c).

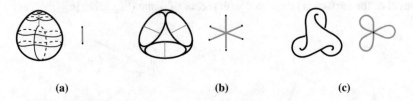

(a)   (b)   (c)

Fig. 6.10 (a) Sphere with a cross-cap   (b) Roman surface   (c) Boy surface

The double point curves of these models of the projective plane are, respectively, a segment with two singular points at the extremities, three non-coplanar segments that intersect at a triple point having six singular points at the extremities, and a rose type curve formed by three non-coplanar petals that intersect at a triple point.

The fourth dimension will be necessary to understand the first two models, while the Boy surface requires a fifth dimension to be fully understood.

6.2   The Sphere with a Cross-Cap

The sphere with a cross-cap is the three-dimensional representation of the projective plane obtained by identifying the boundary of a disk with the boundary of a Möbius strip. To do this, we will transform the boundary of the strip into a curve that closes after one turn.

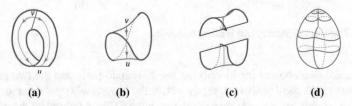

(a)   (b)   (c)   (d)

Fig. 6.11 Building a sphere with a cross-cap

Consider the closed central curve of the Möbius strip, which is called *the soul* of the strip. Choose two diametrically opposite points u and v of the soul. Then, identify the points of the soul that are equidistant from u and therefore equidistant from v (Fig. 6.11a). This identification transforms the soul into a segment having u and v as extremities, and the Möbius strip is transformed into a surface called Plücker's conoid (Fig. 6.11b), after the German mathematician Julius Plücker (1801–1868), Felix Klein's doctoral advisor. The conoid is one of the two surfaces we obtained above when we identified two Whitney umbrellas by the X-shaped part of their boundaries (Fig. 6.11c). The boundary of the Plücker's conoid closes after one turn, and is therefore ready to be glued to the boundary of a disk. The result of this collage is the sphere with a cross-cap (Fig. 6.11d), and the two endpoints of the self-intersecting segment are singular points.

Identifying the soul points that resulted in the self-intersecting segment takes place in the fourth dimension. To see this, we will embed the Möbius strip in four-dimensional space so that its soul is contained in a plane that will be projected onto a straight line in three-dimensional space. This is how the identification of points equidistant from u and v occur: the circular soul is projected onto a segment that has u and v as endpoints.

We will illustrate this process with just one half of the Möbius strip embedded in a hypercube, which delimits a four-dimensional place. The hypercube is projected onto a cube, and half of the soul is projected onto a segment with one singular point at one extremity.

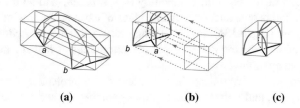

(a)   **(b)**   **(c)**

Fig. 6.12   Projecting half of a Möbius strip from 4D to 3D

Thus, half of the strip embedded in four-dimensional space is projected onto a Whitney umbrella in three-dimensional space. The singular point is exactly the point at which the soul is tangent to the direction of projection. Each of the other points on the self-intersecting segment correspond to two points of the soul in four-dimensional space.

In Fig. 6.12a we have half of the Möbius strip appropriately embedded in the hypercube; that is, a four-dimensional place. The strip is twisted in such a way that only one point of the strip is tangent to the direction of projection from the hypercube to the cube. Figure 6.12b shows the projection to three-dimensional space and Fig. 6.12c shows the resulting Whitney umbrella.

In addition to the convincing geometric representation, the existence of two points of tangency in the projection of the Möbius strip from the fourth to the third

dimension is guaranteed by a particular case of a result, known as the Whitney
conjecture, demonstrated by W. S. Massey (1920–2017) [7].

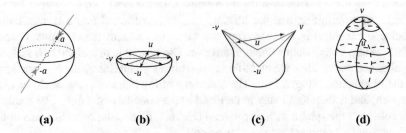

Fig. 6.13 Identifying antipodal points of a sphere

Another way to obtain the sphere with a cross-cap is by identifying the pairs of
antipodal points on a sphere (Fig. 6.13a). The antipodal points are obtained as the
intersection of the sphere with the straight lines that pass through its center.

To identify these pairs of points, we divide the sphere into two hemispheres
and reserve the circle common for both (equator). Having a reflection of the
upper hemisphere relative to the equator, followed by a 180° turn, we obtain the
identification of all antipodal points on the sphere, denoted by a and $-a$, except
the points on the equator, which have been reserved. Next, we identify all the
antipodal points of the equator, denoted by u, $-u$, v, $-v$, and so on (Fig. 6.13b).
This identification is equivalent to sewing edges of extremes u, v with $-u$, $-v$ and
$-u$, v with u, $-v$ (Fig. 6.13c). In this sewing process, a self-intersecting segment is
created, as well as two singular points at the extremities of that segment and both
points are Whitney umbrellas.

The sphere with a cross-cap is a model of the projective plane in three-
dimensional space, and thus any closed non-orientable surface can be represented
in three-dimensional space as connected sum of copies of it.

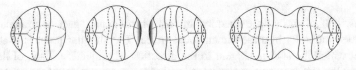

Fig. 6.14 Connected sum of two spheres with a cross-cap

The Klein bottle is the connected sum of two projective planes, $K^2 = P^2 \# P^2$.
Therefore, with two copies of the sphere with a cross-cap, we obtain a model of
the Klein bottle in three-dimensional space (Fig. 6.14). Unlike the traditional Klein
bottle model that has a self-intersecting circle, this model has two self-intersecting
segments and four singular points.

6.3   The Steiner Roman Surface

The Swiss mathematician Jakob Steiner (1796–1863), famous for his contributions to geometry, once on vacation in Rome, created a model of the projective plane, which he called the *Roman surface*.

Fig. 6.15 Roman surface

Similar to the sphere with a cross-cap, this representation of the projective plane in three-dimensional space is also obtained from a Möbius strip. However, in addition to identifying the equidistant points of the soul, as the one made to obtain the sphere with a cross-cap, two other curves are identified in the same way, points equidistant from two chosen points. These other two curves are, similar to the soul of the strip, disorienting paths. The three curves chosen must intersect at three points (Fig. 6.15).

Each disorienting path, after identifying equidistant points, will give rise to a Plücker conoid. Each conoid has a self-intersecting segment with two singular points at the extremities, which are the points chosen to identify the equidistant ones. After identifying this, a Whitney umbrella will appear at each of the two chosen points.

Properly choosing points that define the equidistant identification, the three segments of the three conoids will intersect at a triple point.

The result is Steiner's beautiful Roman surface, which features a triple point and six Whitney umbrellas.

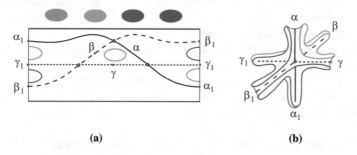

(a)   (b)

Fig. 6.16 Three conoids

To see how this happens, we represent the Möbius strip by its planar model and draw the three disorienting paths on it (Fig. 6.16a).

We choose points α, β and γ, as well as their respective antipodes α_1, β_1 and γ_1. To facilitate viewing, we removed three disks from the Möbius strip, which will be pasted back at the end of the construction.

Identifying the equidistant points on each of the three disorienting paths, from α to α_1, from β to β_1 and from γ to γ_1 yields a composition of three Plücker conoids, whose self-intersection segments intersect at a triple point.

Our representation of this procedure (Fig. 6.16b) is a copy of a drawing made by my colleague, professor J. Scott Carter, and it appears in his interesting book on surface topology [3].

Finally, we glue the three disks that were initially removed and one more disk along the original boundary of the Möbius strip. The result is a model of the Roman surface with its triple point and six singular points.

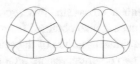

Fig. 6.17 Connected sum of two Roman surfaces

Using two copies of the Roman surface model we obtain another model of the Klein bottle, via connected sum (Fig. 6.17). In this case, the model will have two triple points and 12 singular points.

6.4 The Boy Surface

The two models of the projective plane in three-dimensional space; namely, the sphere with a cross-cap and the Roman surface, both have singular points. These models with transversal self-intersections along lines and a finite number of triple points and Whitney umbrellas are referred to as *semi-regular* models, following usage of the term by Whitney [11].

In this section we will present another model of the projetive plane in three-dimensional space called Boy surface, after the German mathematician Werner Boy (1879–1914), David Hilbert's doctoral student at the University of Göttingen.

In his 1901 thesis, Boy presented a model of the projective plane in three-dimensional space without singular points, with only a single triple point. As we use the term embedding for models without self-intersection, we use the term *immersion* for models with transversal self-intersection but without singular points. Hence, Boy surface is an immersion of the projective plane in three-dimensional space.

François Apéry found a parametrization and also an algebraic degree six defining equation for the Boy surface [1].

Allegedly Hilbert suggested to Boy to prove that every representation of the projective plane in three-dimensional space should have at least two singular points. However, the Boy surface is a counterexample to this supposed Hilbert conjecture.

In the preface of Apéry's book, Egbert Brieskorn (1936–2013) wrote: ... *the famous mathematician David Hilbert was mistaken.*

How could it be that a mathematician as brilliant as David Hilbert proposed a conjecture that his student would annihilate?

In Sect. 6.2, when constructing the sphere with a cross-cap, every projection of an embedding of the Möbius strip from the fourth to the third dimension must have at least two tangency points (Fig. 6.12a), which become singular points in the model in three-dimensional space. This could lead one to thinking that at least two singular points will appear in any representation of the projective plane in three-dimensional space. In fact, this is correct if we restrict ourselves to projections of the projective plane from the fourth to the third dimension. As the projective plane contains a Möbius strip, at least two singular points must appear in such a projection.

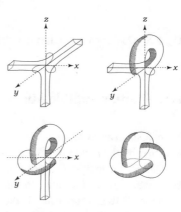

Fig. 6.18 Boy models

Therefore, what Boy managed to do was to avoid all possible points of tangency, embedding the Möbius strip into a five-dimensional space and projecting it from that dimension onto the three-dimensional space. In the space of dimension greater than four, Boy found room for maneuver to bend and twist the Möbius strip and choose a projection that avoided points of tangency, which necessarily occur in the projections of the Möbius strip from the fourth to the third dimension.

Boy's 1904 article in the Mathematische Annalen [2] is the only published work by him. He was killed in combat in World War I, in September 1914. His article contains detailed drawings of the surface construction (Fig. 6.18).

Figure 6.19 shows a sequence of images of a deformation of a Möbius strip creating a triple point and turning the boundary into a circle.

Fig. 6.19 Deforming a Möbius strip to a Boy surface with a hole

The connected sum of two copies of the Boy surface provides another model for the Klein bottle (Fig. 6.20). In this case, the model will have two triple points and no singular points.

Fig. 6.20 Connected sum of two Boy surfaces

6.5   Non-orientable Surfaces in Other Areas

Specialists in areas of knowledge as distinct as psychoanalysis and architecture were attracted by the inside-outside quest associated to non-orientable surfaces.

The psychoanalyst Jacques Lacan (1901–1981) sought in topology, more than a language, a metaphor between the real (outside) and the imaginary (inside). Lacan describes certain properties of the cross-cap that became much admired by some psychoanalysts but completely unknown to topologists.

The cross-cap, or more precisely, the projective plane, can represent the subject of desire in relation to the lost object. A double loop drawn on its surface in effect divides this single-sided surface into two heterogeneous parts: a Möbius strip representing the subject and a disk representing object **a***, the cause of desire. The disk is centered on a point that is related to the irreducible singularity of this surface, which Lacan identified with the phallus* [5].

(a)   (b)

Fig. 6.21 (a) Möbius building by Carlo Séquin   (b) Rem Koolhaas

Some architects have designed buildings using non-orientable surfaces as a model. Peter Eisenman's 1992 Max Reinhardt Haus project (not built) was based on the Möbius strip [4].

Ben van Berkel's Möbius House [10], built between 1993 and 1998 in Het Gooi, near Amsterdam, is also inspired by the non-orientable strip.

Berkeley computer science professor Carlo Séquin's designs are much simpler. He made models of buildings, which were actually based on the Möbius strip (Fig. 6.21a) [8]. In 2012, architects Rem Koolhaas, Cecil Balmond and Ole Scheeren narrowly missed replicating Carlo Séquin's Möbius strip when constructing the China Central Television Headquarters building in Beijing (Fig. 6.21b).

More on the topology-architecture relationship can be found in the inspiring work by David Sperling [9].

References

1. Apéry, François; *Models of the Real Projective Plane*, Vieweg Verlag 1987.
2. Boy, Werner; *Über die Curvatura integra und die Topologie geschlossener Flächen*, Mathematische Annalen 57, 151–184 (1903)
3. Carter, J. Scott; *How surfaces intersect in space*, World Scientific, 2nd edition 1995.
4. Eisenman, Peter; *The Max Reinhardt Haus*, https://eisenmanarchitects.com/The-Max-Reinhardt-Haus-1992
5. Lacan, Jacques; https://nosubject.com/Cross-cap
6. Marar, W. and Mond, D.; *Cover image of the Notices of AMS 44 (3)*, rendering by Thomas Banchoff and Davide Cervone, March 1997. https://www.ams.org/journals/notices/199703/199703FullIssue.pdf?cat=fullissue&trk=fullissue199703
7. Massey, William; *Proof of a conjecture of Whitney*, Pacific Journal of Mathematics 31 (1969).
8. Séquin, Carlo; *-To Build a Twisted Bridge -*, in Bridges, ed. Reza Sarhangi, 23–34 (2000). https://archive.bridgesmathart.org/2000/bridges2000-23.html
9. Sperling, David; *Arquiteturas contínuas e topologia: similaridades em processo*, University of São Paulo Master's dissertation 2003. https://teses.usp.br/teses/disponiveis/18/18131/tde-28032006-155803/en.php
10. Van Berkel, Ben; *Möbius House* (1997) , https://en.wikiarquitectura.com/building/moebius-house/
11. Whitney, Hassler; *The singularities of a smooth n-manifold in (2n-1)-space,* Annals of Mathematics 45, 247–293 (1944).

Chapter 7
Hypersurfaces

7.1 From Surfaces to Hypersurfaces

Geometric objects belong to the Platonic world of ideas, but the objects around us can be used to represent three-dimensional geometric objects, though not all. Objects in our physical world have boundary, which is exactly the part we can see or touch; that is, their surface. For three-dimensional objects finite in size and without boundary, there are no objects in nature that can represent them. Indeed, these objects require at least a four-dimensional environment to embed, and hence they transcend our ability to visualize. We will call such an object a *hypersurface,* although the term is commonly used to designate n-dimensional objects in a $(n+1)$-dimensional environment.

There are some hypersurfaces that are simple generalizations of closed surfaces, and as such it is not difficult to imagine them. The hypercube, for example, is the three-dimensional analogue of the cube. While the cube faces are six solid squares, the hypercube is an appropriated composition of eight solid cubes.

We will call hyperpolyhedra the three-dimensional generalizations of polyhedra. The term polytopes is also used. In addition to the cube, the other four Platonic polyhedra also have three-dimensional analogues; they are regular hyperpolyhedra. These are among the simplest hypersurfaces, along with the hypersphere S^3, which is the three-dimensional generalization of the sphere S^2, and the 3-torus T^3, which is the three-dimensional analogue of the torus T^2.

Speculations about the shape of our universe have awakened astrophysicists' interest in hypersurfaces. Einstein was one of those who believed that understanding the geometric structure of our vast and unknown universe could be facilitated by the knowledge of hypersurfaces. Recognizing the shape of the universe is just the beginning of its full understanding.

We can become acquainted with hypersurfaces through techniques analogous to those used to understand lower-dimensional spaces, such as in the description of the two-dimensional world of Pacmen (Fig. 7.1).

© The Author(s), under exclusive license to Springer Nature Switzerland AG 2022
T. Marar, *A Ludic Journey into Geometric Topology*,
https://doi.org/10.1007/978-3-031-07442-4_7

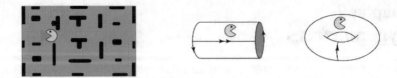

Fig. 7.1 Pacman's world

The Pacmen, tireless explorers, provided us with relevant physical information, with which we created a planar model of their two-dimensional world and deduced its toric shape. Note that Pacman's universe divides our three-dimensional world into two regions, interior and exterior, to which they do not have access.

Other phenomena that take place on the torus can surprise us. Indeed, when faced with a model of the torus in three-dimensional space, a situation that allows us to observe the surface from all angles, certain unexpected geometric facts occur. For example, while in the toric world, there are some paths in which Pacman always moves in the same direction and returns to the starting point. There are many others that, even though Pacman always moves forward in them, will never bring him back to the starting point. Similar conclusions are reached when considering a light beam trajectory in a toric space.

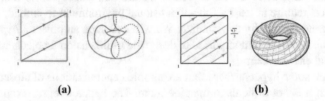

(a)   (b)

Fig. 7.2 Trajectories on a torus

In fact, the curve on the torus whose representation, using its square planar model, consists of straight segments with inclination equals to $1/2$ with respect to the horizontal (Fig. 7.2a) winds twice and closes. Therefore, moving along this curve, after some time you return to the starting point. This is not the case if the slope relative to the horizontal is an irrational number; e.g., $\sqrt{2}/2$ (Fig. 7.2b). In fact, all lines of irrational slopes can wrap infinite times around the torus, and never close.

A world in which you can travel straight ahead and never return to your starting point could lead to the wrong conclusion that it is an unlimited world.

Furthermore, certain intersections of a torus with a set of parallel planes, which result in a set of curves, are surprising. This set of curves, one in each of the parallel planes, is called *tomography,* from the Greek *tomos* that means slice. Widely used in medicine, it serves for reconstructing surfaces using slices obtained by sets of parallel planes, in different directions.

The topological classification of closed surfaces, as seen in Chap. 4, helps to find the shape of the tomographed surface, as the slices obtained from different directions reduce the possible shapes of the list of all closed surfaces. For example, in the case of the sphere S^2, any intersection with a plane is a circle. Therefore, any tomography of the sphere is a set of circles of different radii, whatever the direction of the parallel planes; quite different from the tomography of the torus, which provide a variety of sets of curves, for different directions of the sets of parallel planes.

Indeed, imagine a torus on a horizontal plane. The cuts by planes perpendicular to the base plane provide several types of curves, including one in the shape of the number 8, called *lemniscate* (Fig. 7.3). Cuts through oblique planes may surprise us with two intersecting circles, called Villarceau circles, in honor of the French astronomer Yvon Villarceau (1813–1883).

Fig. 7.3 Torus slices

Adaptation of the tomography to higher dimensional spaces can help us understand some hypersurfaces. While in the case of surfaces (two-dimensional objects), the slices are curves (one-dimensional objects) in parallel planes, in the case of hypersurfaces (three-dimensional objects), the slices are surfaces obtained by cutting through three-dimensional parallel spaces.

Note that just as there are parallel lines in planes and parallel planes in three-dimensional space, in the fourth dimension there are three-dimensional parallel spaces.

Three-dimensional tomography to understand hypersurface shapes will also benefit, in the very near future, from the topological classification of closed three-dimensional objects, which as of today is not yet completed.

The technique used in Chap. 5 to obtain the unfolding of the cube (Fig. 5.14), and adapted to create the open hypercube (Fig. 5.17), can be used to make three-dimensional models of other hypersurfaces.

In fact, in the process of topological classification of closed surfaces, we obtain planar models for each one of them; through cuts, we generate plane polygonal regions with an even number of edges, two for each cut. Similarly, certain hypersurfaces can be represented in three-dimensional space by polyhedral solids with an even number of faces, two for each cut. As a result, three-dimensional representations of hypersurfaces can provide a more simplified geometric understanding of these objects.

In addition to being a powerful technique for understanding the shape of surfaces, planar models can be manipulated when identifying the pair of edges, and reconstruction results in another surface distinct from the original. For example, when identifying the vertical edges of the flat world of Pacman, by turning one of the edges half a turn (180°), we obtain after identification, not a cylinder, but a Möbius strip (Fig. 7.4).

The inhabitants of a flat world shaped like a Möbius strip can travel straight ahead, avoiding the edge, and return to their starting point. However, if the path is

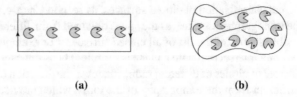

| (a)   (b) |

Fig. 7.4 Pacman on a Möbius strip

over a disorienting path, as seen in Chap. 6, the traveller will return upside down, or rather, as if they had entered a mirror, as their world is two-dimensional (Fig. 7.5).

This transformation, which takes place moving along the Möbius strip, could not be carried out on the surface of a cylinder. In order for an inhabitant of the cylindrical two-dimensional world to become its mirror image, it requires a three-dimensional space in which it will be rotated. Similarly, three-dimensional models of hypersurfaces can also be manipulated to obtain representations of distinct hypersurfaces from the original one.

Fig. 7.5 Mirror image

7.2   On the Shape of the Universe

Our experience in the universe suggests its three-dimensionality, but what would the shape of this three-dimensional universe be like? Would it be finite or infinite in size? This debate has been around for a long time and it is far from a conclusion. This is very good, because in mathematics as in philosophy, as Woody Allen teaches us, a good question is more important than a good answer!

The Greek philosopher Archytas (428 BC–347 BC) was one of the first to record this problem. According to him, the universe has to be infinite, because otherwise, when we reach the end of the world we could stretch out an arm and the world would have to be expanded, indefinitely.

In his 1888 work *L'atmosphère: météorologie populaire* [2], the astronomer Camille Flammarion (1842–1925) published a pleasant wood engraved image in which a missionary of the Middle Ages seems to cross the frontier of the universe (Fig. 7.6).

Archytas assumed the existence of an end of the world, of a boundary. He ignored the possibility of a three-dimensional universe finite in size and without boundary, so his argument does not hold up. In fact, in a finite and without boundary three-dimensional universe; that is a hypersurface, one can always travel

Fig. 7.6 Flammarion 1888

in the same direction and return to the starting point, without ever stretching one's arm out.

In January 1921, Einstein published the article *Geometrie und Erfahrung,* [1] (Fig. 7.7), translated as *Geometry and Experience*, in which he reported something that the general theory of relativity teaches us about the geometric nature of the universe. Einstein presented two possibilities, depending on the average density of matter in the universe. For the hypothesis of an infinite universe to be confirmed, such density needs to tend to zero, otherwise the universe will be finite in size. Einstein assured us that the smaller the mean density, the greater the volume of universal space.

Fig. 7.7 Einstein 1921

Although Einstein suspected that we will never be able to measure the density of matter in the universe just by observations, as it is currently done, he considered the possibility of a three-dimensional universe finite in size and without boundary. In this case, he reported a concern that we mentioned earlier; namely, the visualization of hypersurfaces. In his article, Einstein asked: *Can we picture to ourselves a three-dimensional universe which is finite, yet unbounded?* And he answered himself: *The usual answer to this question is "No," but that is not the right answer. The purpose of the following remarks is to show that the answer should be "Yes." I want to show that without any extraordinary difficulty we can illustrate the theory of a finite universe by means of a mental image to which, with some practice, we shall soon grow accustomed.*

Unfortunately, in his article, Einstein did not provide examples for the practice of such mental imagery.

To test the finite universe hypothesis, it would be interesting to have a list of all possible hypersurface shapes. Once in possession of this list, and backed up with some physical evidence, or with experience in Einstein's words, we will be able to exclude certain formats, and come closer to the solution of this great mystery.

Optimistically, Einstein said it would not take long for astronomers to resolve this issue. However, it is still not resolved. Is a hundred years a long time?

Since the end of the twentieth century, we have witnessed a revolution in the studies of astronomy and astrophysics, accompanied by exponential growth in the amount of information gathered. This was accentuated by the launch of the Hubble telescope (Fig. 7.8) on April 24, 1990 aboard the Discovery spacecraft. The next day, orbiting our planet, Hubble started observing the universe, without the distortions that telescopes on Earth are subject to. In 2004, Hubble was pointed to

Fig. 7.8 Hubble

a region of the Cosmos that was not yet known, southwest of Orion and near the constellation Fornax. After a few days, astronomers had a huge surprise: the photographic record of that adventure contained about ten thousand galaxies, each with billions of stars. The region came to be known as the Ultra Deep Field and is the deepest image of the universe taken in visible light, a record of more than 13 billion years ago. In fact, as the speed of light is finite, reflected or emitted light takes a while to reach our retina; therefore, everything we see is an image of the past!

The Hubble telescope helped to confirm that we still know little about our universe and that this is a big place, so big that, according to some, it would be a huge waste if only our planet were inhabited.

On February 22, 2017, NASA announced a spectacular discovery of a cluster of seven planets, not far from here, that orbit a star similar to our Sun and are believed to be habitable.

A pale blue dot was how the physicist Carl Sagan (1934–1996) described the planet Earth as seen from Saturn, in the photographic record sent by the Voyager spacecraft in 1990. The Earth is a mere point in a vast universe, as some philosophers had already claimed in Ancient Greece.

It is believed that since ancient times, humanity has sought to understand the Cosmos, and some archaeological monuments support this. For example, some believe that Stonehenge in England (Fig. 7.9) was built over 5 thousand years ago as a kind of astronomical observatory. According to British archaeologist Evan Hadingham, Stonehenge was not used by stone age Einsteins, but rudimentary barbarians preoccupied with rituals involving life, death, fertility and their ancestors.

Fig. 7.9 Stonehenge

Even after many studies have been carried out, our universe remains mysterious and unsettling. However, new discoveries, both in astronomy and in geometry, show that we are expanding our knowledge, albeit slowly.

The research results obtained by the Bulgarian mathematician and astrophysicist Fritz Zwicky (18980–1974) are exemplary. Zwicky knew that a group of rotating galaxies held together by gravitational force; otherwise, depending on the speed, they would be thrown off like people on a speedy merry-go-round if they do not hang on.

In 1933, during his observations Zwicky found an incongruity: a group of galaxies whose gravitational force that held them together turned out to be greater than the force caused by the observed matter. The gravitational force was there, but you could not see the matter corresponding to that force. Thus, Zwicky concluded that there must be unobservable matter; that is, matter that does not reflect light, which he called *dark matter*.

Initially Fritz Zwicky's arguments were not well accepted by the scientific community. Only in 1970, after sufficiently accurate measurements were taken by astronomer Vera Rubin (1928–2016), did the idea of dark matter become accepted. However, the beginning of Rubin's scientific career was not an easy one. In 1948, after earning a degree in Astronomy from Vassar Female College, she applied for a doctorate at Princeton University, but her application was rejected. Princeton did not accept women to study Astronomy until 1975.

Although physicists have not yet detected dark matter, they believe they have an idea of how much of it exists in the universe based on observations of galaxies. Some even believe that dark matter lives in a fourth dimension.

Current data suggest that only 4% of the universe is made up of baryonic matter (composed mainly of protons, neutrons and electrons) that reflects light. Could it be more uncanny? Yes, because of the remaining 96% of the universe, 3/4 would be in the still mysterious form of dark energy and only 1/4 is dark matter.

Given the difficulty to understand the composition of matter in the universe, as well as its shape, some scientists have concluded that the two problems are related. Many researchers believe in the surreal hypothesis of an infinite universe, while some, such as Einstein, accept the possibility of a three-dimensional model finite in size and without boundary, therefore a hypersurface.

Some ability to visualize hypersurfaces will be necessary to perceive the possible shape of our universe, but it is not sufficient. This is a difficult problem, as many others we have faced. Consider that even the shape of our planet was the subject of much dispute in the early days; several possibilities were considered, until reaching the spherical model. In Christopher Columbus' time, around the year 1490, the hypothesis circulated that our planet was pear shaped. On a microscopical scale, according to a recent article by L. P. Gaffney et al. [3], the pear shape in the nucleus of certain atoms was unexpectedly identified, and this has called into question certain traditional theories of physics.

When Russian astronaut Yuri Gagarin (1934–1968) was launched into space in 1961 on the Vostok 1 spacecraft, he saw not only the shape but also the color of our planet.

The geometric perception of the universe in which we live is very complicated. It is often easier to appreciate something when you are outside of it.

In the film *Jeder für sich und Gott gegen alle* (translated as *The Enigma of Kaspar Hauser*) by German director Werner Herzog, the character Kasper Hauser, who since birth had been trapped for 17 years in a room inside a tower, when he is first taken out of the tower, makes the following observation about the nature of space: *In my room, looking left, right, forward, backward, up and down, I can only see my room. Out here, I look at the tower, but if I turn around, the tower disappears. So, my room is bigger than the tower.* The small room inside the tower, where Kasper had always been confined, was existentially his entire universe.

This beautiful blue planet of ours (at least from a distance, as it is increasingly polluted and gray up close) is part of a solar system, which is part of the Milky Way, a galaxy, which in turn is part of the local Group, which is a collection of several galaxies, which in turn .... We live in a very complex universe and there is no chance of leaving it to enjoy its shape, as Gagarin did with our planet. We will therefore have to deduce this shape, and mathematics can help us.

According to Einstein, one reason why mathematics enjoys special esteem, above all other sciences, is that its laws are absolutely certain and indisputable, while those of all other sciences are to some extent debatable and in constant danger of being overturned by the discovery of new facts.

7.3   Three-Dimensional Objects

Since there is no three-dimensional objects in our physical world that can model hypersurfaces, let us try to understand them by analogy, based on the familiarity we have with closed two-dimensional objects, which precisely represent the boundaries of three-dimensional objects in our physical world. This analogy will be facilitated with a good understanding of some concepts that are registered as definitions in the first of the 13 books of Euclid's Elements.

Definition 13 *A boundary is that which is an extremity of anything.*

The following definition describes the set of points interior to a boundary:

Definition 14 *A figure is that which is contained by any boundary or boundaries.*

These are quite vague definitions and require some examples for clarification.

Example 1 A line segment has two endpoints. The interior points of the segment constitute a figure. We will say that the line segment is a figure that has two boundary points.

In a plane, the region contained by a circle is the figure called *disk*. Disks are two-dimensional figures with one-dimensional circular boundaries.

In three-dimensional space, the region contained by a sphere is the figure called *solid sphere*. The solid sphere is a three-dimensional figure that has the sphere as its two-dimensional boundary.

In topology, a disk is called 2-disk and is denoted by D^2. The boundary of D^2 is called 1-sphere and denoted by S^1. This generalizes to any dimension, an n-disk D^n is an object whose boundary is an $(n-1)$-sphere . Also, the line segment is called 1-disk and its boundary, the two endpoints, is called 0-sphere.

7.3.1   The Hypersphere

Let $\mathbb{R}^n$ denote the Euclidean n-dimensional space. We will define the hypersphere as a subset of the four-dimensional space $\mathbb{R}^4$. The sphere, bidimensional object, is well known as a subset of $\mathbb{R}^3$. If C is a point in $\mathbb{R}^3$ and r is a given length, then the sphere of radius r and center C (Fig. 7.10c), denoted by S^2, is the set of points of $\mathbb{R}^3$ whose distances to point C are all equal to r, it is a two-dimensional object without boundary.

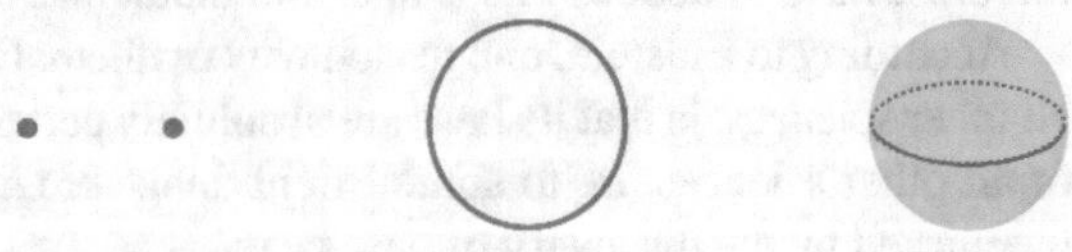

Fig. 7.10 Zero-dimensional, one-dimensional and two-dimensional spheres

(a)   (b)   (c)

With the same definition, but collecting the points in $\mathbb{R}^2$, we obtain the circle of radius r and center C (Fig. 7.10b), the S^1, or sphere of dimension 1, a one-dimensional object without boundary. Collecting the points in the line $\mathbb{R}$, following the same definition, we obtain only two points that are distant r from the point C, this is the S^0 (Fig. 7.10a), or sphere of dimension 0.

Similarly, we define the hypersphere, denoted by S^3, as the set of points in $\mathbb{R}^4$, equidistant (radius) from a fixed point (center). Likewise, the n-dimensional sphere is defined as a subset of $(n + 1)$-dimensional space.

A mental image of the hypersphere S^3 can be constructed using a three-dimensional tomography. While any tomography of the sphere S^2, no matter the direction of the parallel planes, always produce circles S^1 whose radii increase until reaching a maximum equal to the radius of the sphere; in the hypersphere S^3 the intersections by parallel three-dimensional spaces produce spheres S^2 whose radii increase until reaching a maximum equal to the radius of the hypersphere.

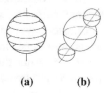

(a)   (b)

Fig. 7.11 Sphere and hypersphere

In fact, let us consider on the sphere (a two-dimensional object without boundary) two antipodal points, the north pole and the south pole. Slices of the sphere by planes (two-dimensional spaces) perpendicular to the axis determined by the poles provide circles (one-dimensional objects without boundary) with centers at the axis points and radii that increase to maximum at the equator, then decrease to zero at the poles (Fig. 7.11a). Analogously, we distinguish two points on the hypersphere and call them the N pole and the S pole. Cuts of the hypersphere by three-dimensional spaces perpendicular to the axis from the N pole to the S pole, provide spheres (bidimensional objects without boundary) with centers at points on that axis and radii that increase until reaching a maximum, and then decrease until zero. The sphere is thus perceived as a collection of stacked circles, with centers along the axis determined by the north and south poles. Likewise, a hypersphere is a collection of stacked spheres with centers along the axis determined by the N and S poles (Fig. 7.11b).

The hypersphere S^3 can also be obtained geometrically by identifying two solid spheres by their boundary. Recall that a solid sphere is a three-dimensional object and its boundary is a two-dimensional sphere S^2. To understand the identification of the boundary points of two solid spheres, let us first visualize the analogous process to obtain spheres S^1 and S^2.

As we have seen, the set of points on the line that are interior to a S^0 is the line segment D^1, the 1-disk. Having two 1-disks, we can create a circle (topologically) identifying, point by point, the points on the boundary of one figure with those of the other. Let us deform the two segments into arcs, bringing the boundary points of one to the boundary points of the other, until they are glued together. Physically, we imagine that our figures are made of a perfectly deformable material and so we create a physical model of a circle from two segments. After identifying them, the boundary points disappear, and therefore circle S^1 is a one-dimensional object without boundary (Fig. 7.12a).

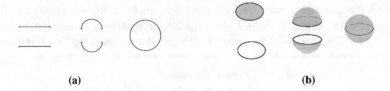

(a)   **(b)**

Fig. 7.12 (a) One-dimensional sphere   (b) Two-dimensional sphere

We will do the same with two-dimensional objects. The set of points on the plane that are interior to circle S^1 is the 2-disk D^2.

Consider two 2-disks with their circular boundary. By identifying the corresponding points of one circular boundary with those of the other, we obtain, topologically, sphere S^2. After this identification the boundary points disappear and, therefore, sphere S^2 is a two-dimensional object without boundary (Fig. 7.12b).

Finally, we will do the same with three-dimensional objects. Consider the two solid spheres; that is, two 3-disks D^3 whose boundaries are spheres S^2. Identifying the corresponding points of one spherical boundary with those of the other, we obtain, topologically, hypersphere S^3.

In summary, two 1-disks have their boundary S^0 identified point by point, giving rise to S^1. Two 2-disks have their boundary S^1 identified point by point, giving rise to S^2. Two 3-disks have their boundary S^2 identified point by point, giving rise to hypersurface S^3.

While the procedures for obtaining S^1 and S^2 are easy to visualize, the point-by-point identification of boundary points of two 3-disks requires a certain abstraction. In fact, the hypersphere lives in a fourth dimension, so a three-dimensional physical experience is not sufficient for that visualization.

There is an alternative way of conceiving the hypersphere, through the concept of *suspension of a set.*

Let S be a set of points. Let P_1 and P_2 be two points not belonging to S. We call suspension of S the set of all line segments with one end at P_1 and the other at each point of S, together with the set of all line segments with one end at P_2 and the other at each point of S. Points P_1 and P_2 are called suspension vertices.

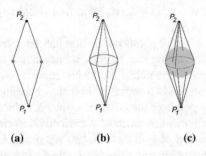

              (a)   **(b)**   **(c)**

Fig. 7.13 Suspension of spheres

Therefore, the suspension of S^0 is, topologically, S^1 (Fig. 7.13a). Similarly, the suspension of S^1 is S^2 (Fig. 7.13b), and the suspension of S^2 is hypersphere S^3.

Even as a suspension of S^2, the hypersphere is not easy to imagine.

However, the two vertices P_1 and P_2 of the suspension can be chosen randomly, as long as they do not belong to the set to be suspended. Thus, choosing the suspension vertices properly can help in the visualization. In the case of suspension of S^2 to generate S^3, we can choose vertex P_1 at the center of S^2, so P_1 is inside S^2 and the other vertex P_2 is chosen outside S^2.

Having chosen this, the suspension of S^2 is easier to see. In fact, the set of straight line segments starting at P_1, the center of S^2, and following to the points of sphere S^2 constitutes a solid sphere D^3, whose center is in P_1. Furthermore, the set of segments that start at the outer point P_2 and follow to the points of sphere S^2 constitutes, topologically, a solid sphere. These two solid spheres have a common boundary, the points of the suspended S^2.

Fig. 7.14 Dante's universe

In his 1979 article *Dante and the 3-sphere* [5], mathematical physicist Mark A. Peterson suggests that hypersphere S^3 coincides with the universe described by Dante Alighieri (1265–1321) in *La Divina Commedia*. The suspension of S^2, with an inner and an outer vertex, with God in the outer luminous point and Satan in the dark inner point (Fig. 7.14), describes, according to Peterson, Dante's universe in Canto 28 of Paradise.

Another way to visualize hypersphere S^3 is through its hyperpolyhedral representation: the hypercube.

In fact, the cubic figure; that is, the solid cube, is a polyhedral representation of the solid sphere. Identifying two solid cubes by their cubic boundary generates a hypercube, which is a hyperpolyhedral representation of the hypersphere. Topologically, the hypersphere and the hypercube coincide.

A well-known representation of the hypercube as a projection from four-dimensional to three-dimensional space and down to the plane of this page is shown in Fig. 7.15. It is generated by identifying the boundaries of two solid cubes, but the image can be quite deceitful because one of the solid cubes is placed inside the other solid cube.

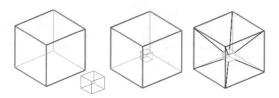

Fig. 7.15 Hypercube

Next, we will show other examples using the Euclidean definition 14 of figures to support the configuration of other hypersurfaces.

Example 2 Three non-collinear points define a plane. The occurrence of three collinear points is something very particular, while three non-collinear points is a more common phenomenon. We will say that three points are in *general position* if they are not collinear. Likewise, four points are in general position if they are not all on the same plane. Analogously, we define n points in general position. In a plane, the set of three segments whose endpoints are three points in general position defines a triangle (one-dimensional object without boundary). The triangle is the boundary of a triangular figure in the plane. The three points that define the triangle are called *vertices* of the triangular figure, the segments defined by the pairs of points are called *edges*, and the triangular figure is called *face*.

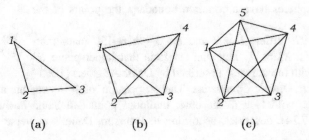

(a)   (b)   (c)

Fig. 7.16 Points in general position

Example 3 Let us consider four points in general position; that is, four non-coplanar points. In principle, it is impossible to represent these four non-coplanar points in the plane of this page, but we pretend that they are in different planes. Unlike traditional drawing techniques such as perspective, in mathematics a drawing represents exactly what we want it to represent! For example, while non-collinear points 1, 2 and 3 are in the plane defined by them (Fig. 7.16a), point 4 must be imagined outside of that plane (Fig. 7.16b).

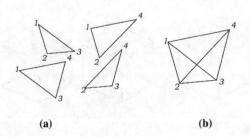

(a)   (b)

Fig. 7.17 Four faces

If P_1, P_2, P_3 and P_4, are four points in general position, then each of the four triples defined by them (P_1, P_2, P_3), (P_1, P_2, P_4), (P_1, P_3, P_4), (P_2, P_3, P_4) defines a triangular figure in three-dimensional space as the four points are non-coplanar.

The set of these four triangular figures is a tetrahedron (Fig. 7.17b). It has four vertices, six edges and four faces. According to Euclid, the tetrahedron figure is the set of points in three-dimensional space contained by the tetrahedron. This three-dimensional figure has the tetrahedron (two-dimensional) as its boundary. The tetrahedral figure (three-dimensional) is called a *cell*.

Example 4 Let us consider five points in general position; that is, they do not belong to the same three-dimensional space (Fig. 7.16c). Note that just as there are infinitely many lines in the plane and infinitely many planes in three-dimensional space, there are infinitely many three-dimensional spaces in four-dimensional space.

The five possible subsets of four points, defined by those five points in general position, will be vertices of tetrahedral figures, each in a different three-dimensional space (Fig. 7.18). Thus, five points in general position are vertices that define 10 edges, which in turn define 10 triangular faces, which define five tetrahedral cells. This composition is called 5-cell (Fig. 7.19).

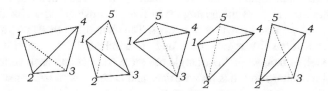

Fig. 7.18 Five cells

While a triangle is the boundary of a triangular figure (plane region) and a tetrahedron is the boundary of a tetrahedron figure (three-dimensional region), the 5-cell is the boundary of a four-dimensional region.

The 5-cell is an example of a hypersurface; that is, a three-dimensional object finite in size and without boundary. We cannot see it physically, as the 5-cell lives in the fourth dimension. The most we can see of the 5-cell in our three-dimensional space is one of its tetrahedral cells. In fact, when a tetrahedral figure is supported on a plane, only one of its four faces (triangular figures) will belong to that plane. Similarly, when a 5-cell is supported in three-dimensional space, only one of its five cells (tetrahedral figures) will appear in three-dimensional space, and that is all that can be seen of a 5-cell.

Fig. 7.19 5-cell

The anguish that may be felt for not being able to visualize the 5-cell is the same that flatlanders suffer (Chap. 5), beings who live in a two-dimensional world, they can only appreciate one of the triangular faces of a tetrahedral figure, because this figure belongs to three-dimensional space; an esoteric world for them.

7.3.2   Regular Hyperpolyhedra

We can form polygons using vertices and edges, one-dimensional objects without boundary. Regular polygons have all edges of the same length and two edges meet at each vertex. A regular polygonal figure, that is, a regular polygon together with its interior points is called a regular face.

We can form polyhedra using vertices, edges and faces, bidimensional objects without boundary. Regular polyhedra are those whose faces are regular polygonal figures all of the same type and size, and the same number of regular faces meet at every vertex. Under these conditions, we saw in Chap. 2 that there are only five regular polyhedra, the Platonic polyhedra. A regular polyhedral figure; that is, a regular polyhedron together with its interior points is called a regular cell. The Platonic solids are the five regular cells.

We can form hyperpolyhedra using vertices, edges, faces and cells; that is, three-dimensional objects without boundary, therefore hypersurfaces. Regular hyperpolyhedra are those whose cells are regular polyhedral figures all of the same type and the same number of regular cells meet at every vertex. Each vertex becomes indistinguishable.

The Swiss mathematician Ludwig Schläfli (1814–1895) described all regular hyperpolyhedra. They are the analogues in four-dimensional space of the regular polyhedra, and are also known as polytopes. The simplest of these polytopes is the regular 5-cell. It has five regular cells that are tetrahedral figures and it is easy to build a model. In fact, while the triangle has three vertices and each vertex has two edges, and the tetrahedron has four vertices and each of them has three edges, the 5-cell has five vertices and each of them have four edges.

The procedure for constructing regular hyperpolyhedra models can be generalized to obtain other polytopes. Let us list the polytopes whose faces are square figures.

The square has 4 vertices ($4 = 2^2$) and at each vertex there are two edges. The square encloses a region of two-dimensional space. The cube has 8 vertices ($8 = 2^3$) and each of them has three edges. The cube encloses a region of three-dimensional space.

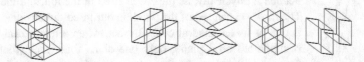

Fig. 7.20 Hypercube's eight cells

The hypercube has 16 vertices ($16 = 2^4$) and each of them has four edges. The hypercube encloses a region of four-dimensional space. Therefore, successively, 2^n is the number of vertices of a region of n-dimensional space and in each vertex there are n edges. Furthermore, such an object has $2n$ hypercells of dimension $n - 1$.

The hypercube is the regular hyperpolyhedron corresponding to the regular polyhedron cube. It is also known as the regular 8-cell because of its eight cubic cells (Fig. 7.20).

Returning to the polytopes whose faces are triangular figures, we also have those corresponding to the octahedron and icosahedron. They are: the 16-cell (Fig. 7.21a), which is the regular hyperpolyhedron corresponding to the regular octahedron, and has 16 cells, which are regular tetrahedron figures, and the 600-cell (Fig. 7.21c), which is the regular hyperpolyhedron corresponding to the icosahedron, and has 600 regular tetrahedron cells.

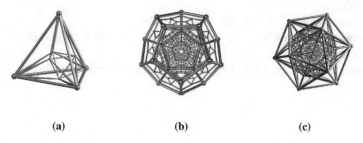

(a)   (b)   (c)

Fig. 7.21 16-cell, 120-cell and 600-cell

The only polytope whose faces are pentagonal figures is the 120-cell (Fig. 7.21b), which is the regular hyperpolyhedron corresponding to the regular dodecahedron, and has 120 dodecahedral cells.

The 5-cell, 8-cell, 16-cell, 120-cell and 600-cell are the four-dimensional analogues to the five regular polyhedra of three-dimensional space, respectively the tetrahedron, cube, octahedron, dodecahedron and icosahedron.

Surprisingly, in addition to these five polytopes, Schläfli discovered one more regular hyperpolyhedron in four-dimensional space. It is the 24-cell (Fig. 7.22), and this one does not have an analogue in three-dimensional space, where only the Platonic five exist. It has 24 cells, which are regular octahedral figures.

Note that the hyperpolyhedra models, although difficult to perceive when drawn on the plane of this page, follow the pattern of construction of polygons and polyhedra. They are the simplest hypersurfaces to draw. The beautiful images of 16-cell (Fig. 7.21a), 24-cell (Fig. 7.22), 120-cell (Fig. 7.21b) and 600-cell (Fig. 7.21c) were drawn with the Stella software by Robert Webb.

Fig. 7.22 24-cell

Three-dimensional objects without boundary require at least the fourth dimension for a possible embedding. We have already experienced the same difficulty with closed non-orientable surfaces (Chap. 6), as they also require a four-dimensional space for an embedding. Therefore, hyperpolyhedra models surpass the limits of our visual capacity.

However, we can overcome this difficulty by
making three-dimensional models analogous to
the planar model of closed surfaces. For example,
a three-dimensional model of the 5-cell can be
obtained through six cuts along the triangular
faces whose vertices, among the five 1 2 3 4 5,
are represented by the triples [123], [124], [125],
[134], [135] and [145]. Its five cells, tetrahedral figures, have vertices given by the quadru-
ples [1234], [1235], [1245], [1345] and [2345]
(Fig. 7.23).

Another example, the 8-cell; that is, the
hypercube is a hypersurface for which a three-
dimensional model is obtained through 17 cuts
along square faces (Fig. 5.17). This three-dimensional model is called open hypercube.

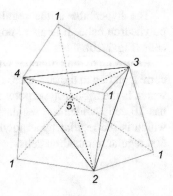

Fig. 7.23 Open 5-cell

7.3.3 The 3-Torus

The 3-torus is the three-dimensional analog of the two-dimensional torus. The 3-
torus, which is denoted by T^3, has a three-dimensional model that follows that of the
torus T^2. Recall that T^2 is represented by a planar model which is a square figure,
with pairs of opposite edges identified. We can see T^2 as a sequence of spheres
S^1 going around another S^1, we write $T^2 = S^1 \times S^1$. Analogously, the 3-torus is
represented by a three-dimensional model in the shape of a solid cube, with pairs
of opposite faces identified. These face identifications are direct, point-to-point,
without rotation, analogous to the edge identifications in the torus reconstruction.
While the identifications of pairs of opposite edges of the square are made in three-
dimensional space, the identifications of pairs of opposite faces of the solid cube
occur in the fourth dimension. We can see T^3 as a sequence of torus T^2 going
around a sphere S^1, we write $T^3 = S^1 \times S^1 \times S^1$.

Just as a torus is a closed surface that divides three-dimensional space into two
regions, one inside and one outside, the 3-torus is a hypersurface that divides four-
dimensional space into two regions. In the same way that Pacman's universe is a
surface of toric shape, our universe could have the shape of a 3-torus. Indeed, in
1984, Russian astrophysicist Alexei Starobinsky proposed a cosmological model
where the shape of the universe is a 3-torus.

Let us see how to represent a walk through the 3-toroidal universe using the three-
dimensional solid cubic model. Consider that the walk happens in four-dimensional
space.

Our closed path starts at a point a, at the base of the cube (Fig. 7.24). Note that
due to the identification of opposite faces, point a is also on the top face. Our path
goes east to the point b and so, going through that face, we reappear at the west face

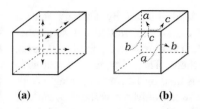

Fig. 7.24   3-torus

point b. Then, we go to point c on the back face and reappear at the same point c on the front face, ending at point a.

Manipulations with planar models give rise to representations of new surfaces. For example, in the square model, rotating an edge of $180°$ before identifying it with the opposite edge, and having performed a direct identification of the other pair of edges, we obtain a Klein bottle instead of a torus. By analogy, we can identify one pair of opposite faces of a solid cube after a $180°$ turn, and the other two pairs of faces with direct identification. This would result in a three-dimensional object without boundary, which can be seen as a sequence of Klein bottles K^2 going around a sphere S^1; that is, $K^2 \times S^1$.

We have seen that an inhabitant of the two-dimensional world in the shape of the Möbius strip, when following a disorienting path, returns to the starting point transformed into its mirror image. What could happen to us if we traveled straight ahead, on a disorienting path in $K^2 \times S^1$ and back to where we started?

More on this subject can be found in the excellent book *The shape of space* by Jeff Weeks [6].

7.4   A Model of the Universe

In the early twentieth century, Einstein showed that the gravitational force generated by massive celestial bodies can change the path of light, affect time and determine the shape of the universe. Based on his general theory of relativity, he proposed a hyperspherical model of the universe with a non-Euclidean geometry.

One hundred years later, Jeff Weeks and Jean-Paul Luminet led groups of mathematicians and astronomers that seek to describe geometric aspects of the universe. Some call this area of mathematical physics *Cosmic Topology.*

The Platonic speculation of a dodecahedral universe found support in the scientific data of these groups, reported in the article by Luminet, J.-P. et al [4].

Data and measurements from the WMAP satellite, which examines microwaves generated shortly after creating the universe, confirm the conjecture of a three-dimensional universe finite in size and without boundary, a hypersurface based on the dodecahedron!

The authors' idea is developed by analogy with a dodecahedral tiling of the sphere. Initially, a dodecahedron is considered, which comprises 12 pentagonal faces, circumscribed by a sphere. Then, the dodecahedron is inflated and the sphere is tiled by 12 curved pentagons. To model our three-dimensional universe without boundary, the authors considered a 120-cell circumscribed by a hypersphere. Then, the 120-cell is inflated and becomes a three-dimensional tiling of the hypersphere. According to the authors, this would be the possible shape of our universe (Fig. 7.25).

Fig. 7.25 Nature 2003

Calculating the diameter of this model, the spectacular distance of approximately 30 billion light years is obtained, in which one light year is equivalent to the distance traveled by light in a vacuum in a Julian year, thus approximately 10 trillion kilometers.

Some articles in the journal Nature, such as the one above, are incredible, to the point that they do not hold up for many years (Fig. 7.26).

In August 2016, was reported on TV the discovery of a planet that is very similar in size to Earth, orbiting the star Proxima Centauri, which is the closest star to our sun. The feat was described in an article, also published in the journal Nature, and is very encouraging.

After doing everything to make our planet more uninhabitable, it seems that a new house is being offered to us, but a little far from here, more than four light years away. It is estimated that nuclear-powered spacecraft, which exist only in theory, would take at least 100 years to reach the new planet.

In November 2012, the same reputable journal published a similar article by eleven European astronomers.

Fig. 7.26 Nature 2016

The discovery of a planet with a mass similar to the Earth's mass, orbiting α Centauri B, was announced. However, this discovery was rejected by some astronomers who found errors in the paper's conclusions. The 2016 article, however, has 31 signatories, which indicates that the discovery will probably hold. The future will tell.

The universe is so immense that it appears immutable, and that the duration of a planet such as that of the earth is only a chapter, less than that, a phrase, less still, only a word of the universe's history. Camille Flammarion, La Fin du Monde, 1894.

References

1. Einstein, Albert ; *Geometrie und erfahrung*, Springer (1921). English translation https://mathshistory.st-andrews.ac.uk/Extras/Einstein_geometry/
2. Flammarion, Camille; *L'atmosphère: météorologie populaire*, Hachette Livre 1888
3. Gaffney, L. P. et al.; *Studies of pear-shaped nuclei using accelerated radioactive beams*, Nature 497, 199–204 (2013).
4. Luminet, J.-P. et al.; *Dodecahedral space topology as an explanation for a weak wide-angle temperature in the cosmic microwave background*, Nature 425, 593–595 (2003).
5. Peterson, Mark; *Dante and the 3-sphere*, American Journal of Physics 47, 1031–1035 (1979).
6. Weeks, Jeffrey; *The shape of space*, CRC press 2001.

Printed in the United States
by Baker & Taylor Publisher Services